T0259768

Birkhäuser

Frontiers in Mathematics

This series is designed to be a repository for up-to-date research results which have been prepared for a wider audience. Graduates and postgraduates as well as scientists will benefit from the latest developments at the research frontiers in mathematics and at the "frontiers" between mathematics and other fields like computer science, physics, biology, economics, finance, etc. All volumes are online available at SpringerLink.

Soon-Mo Jung

Ulam's Conjecture on Invariance of Measure in the Hilbert Cube

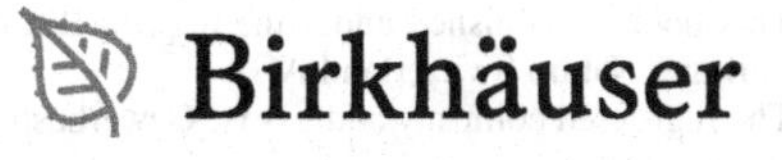

Soon-Mo Jung
Mathematics Section
Hongik University (Sejong Campus)
Sejong, Korea (Republic of)

ISSN 1660-8046 ISSN 1660-8054 (electronic)
Frontiers in Mathematics
ISBN 978-3-031-30885-7 ISBN 978-3-031-30886-4 (eBook)
https://doi.org/10.1007/978-3-031-30886-4

Mathematics Subject Classification: 28A35, 28A78, 46B04, 46C05, 46G12

This book is published under the imprint Birkhäuser, www.birkhauser-science.com by the registered company Springer Nature Switzerland AG
The registered company address is: Gewerbestrasse 11, 6330 Cham, Switzerland

Preface

Ulam's conjecture, which claims the invariance of the standard product probability measure defined on the Borel σ-algebra of the Hilbert cube, was probably made in the late 1930s before the outbreak of World War II. However, Ulam's conjecture did not become known to the world until the following book published by Stanislaw M. Ulam in 1960:

- S. M. Ulam, *A Collection of Mathematical Problems*, Interscience, New York, 1960.

In fact, Ulam conjectured the invariance of measures defined on the compact metric space:

> Let X be a compact metric space. Does there exist a finitely additive measure μ defined for at least all the Borel subsets of X, such that $\mu(X) = 1$, $\mu(p) = 0$ for all points p of X, and such that congruent subsets of X have equal measure?

In the 1970s, J. Mycielski rephrased Ulam's conjecture using modern mathematical terms:

> The standard product probability measure on the Borel σ-algebra of the Hilbert cube is invariant.

The above statement is now widely known as Ulam's conjecture. He also succeeded in proving Ulam's conjecture under the assumption that the sets involved in the conjecture are open.

A few years later, J. W. Fickett, who closely analyzed Mycielski's work, was able to partially prove Ulam's conjecture by transforming the δ-isometry to construct an exact isometry and combining it with the method developed by Mycielski. In fact, he proved Ulam's conjecture for metrics defined by any sequence of positive real numbers that decreases to 0 very quickly.

More than 30 years later, S.-M. Jung and E. Kim partially proved Ulam's conjecture by improving Fickett's method and constructing a Hausdorff measure using the concept of entropy. Their result significantly improved Fickett's result.

In 2020, S.-M. Jung combined various methods to prove Ulam's conjecture. Among them, he developed and applied a method for extending the domain of local isometries, and was able to completely prove that Ulam's conjecture is true.

This book is structured to give the reader an overview of the whole process of solving Ulam's conjecture, with details of the proof process explained in detail. Moreover, since this book is written in an integrated and self-contained manner, it is easy for readers to understand the contents and to avoid the trouble of searching for literature on their own.

This book consists of six chapters, and we will now briefly describe the content of each chapter.

- Chapter 1 briefly introduces the basic concepts and theorems of general topology that are essential to understanding the subject of this book. All lemmas and theorems that are necessary to prove the theorems presented in this chapter are introduced so that the reader does not have to reference other literature when reading this book.

- In Chap. 2, we briefly introduce basic concepts and theorems in vector space, normed space, Banach space, inner product space, and Hilbert space that are essential to prove Ulam's conjecture.

- The concept of measures is a generalization and formalization of length, area, volume, and other common notions such as mass and probability of events. Although these concepts may appear different at first glance, they have many similarities and can often be unified by the concept of measure. In Chap. 3, we briefly present the basic concepts and theorems of general measure theory that are essential for proving Ulam's conjecture.

- In Chap. 4, we define the first- and second-order generalized spans and the index set, examine their properties, and apply them to the study of the extension of isometries. To this end, we develop a theory that extends the domain of local isometries to the generalized spans. We also prove that under the axiom of choice, the domain of a local isometry can be extended to any real Hilbert space, where the domain of a local isometry need not be a convex body or an open set.

- In 1974 and 1977, J. Mycielski published the first papers dealing with Ulam's conjecture. In fact, he proved Ulam's conjecture positive with the additional assumption that the relevant sets are open. In 1982, J. W. Fickett succeeded in partially proving Ulam's conjecture by proving the following statement in a different way than Mycielski: If the a_i's decrease to 0 very rapidly, then any two Borel subsets of the Hilbert cube, which are d_a-isometric, have the same standard product probability measure. Around 40 years later, in 2018, S.-M. Jung and E. Kim jointly studied Ulam's conjecture and improved Fickett's result by partially proving Ulam's conjecture. In Chap. 5, we introduce the historical process of solving Ulam's conjecture by presenting a summary of these papers. In some cases, the reader may skip reading this chapter.

- The conjecture of Ulam states that the standard product probability measure on the Hilbert cube is invariant under the induced metric d_a when the sequence $a = \{a_i\}_{i\in\mathbb{N}}$ of positive real numbers satisfies the condition $\sum_{i=1}^{\infty} a_i^2 < \infty$. In Chap. 6, we completely prove that Ulam's conjecture is true under the axiom of choice.

The author would like to express his sincere thanks to Professor Dr. Dietmar Kahnert for suggesting and guiding this topic as a subject of his doctoral thesis. The author cannot fail to express his sincere thanks to Mr. Doyun Nam of Seoul National University, who has improved the quality of this book by carefully reading the first manuscript of this book and pointing out various points for improvement.

This work was supported by the National Research Foundation of Korea (NRF) grant funded by the Korean government (MSIT) (No. 2020R1F1A1A01049560).

Sejong, Republic of Korea Soon-Mo Jung
February 2023

Contents

Chapter 1

Topology

In this chapter, we will briefly introduce the basic concepts and theorems of general topology necessary to understand the subject matter of this book. Among many other literatures listed in the References section, we mainly refer to the book [13] by R. H. Kasriel and the book [19] by G. F. Simmons for this purpose. All the lemmas necessary to prove the theorems presented in this chapter are introduced so that the reader, if possible, does not have to refer to other books when reading this book.

1.1 Basic Concepts for Metric Spaces

A metric space is a set along with a metric on the set, and the metric is a function that generalizes the notion of a distance between any two elements of a set. In other literature, elements of sets are often referred to as points, but this book will refer to them as elements instead of using the term points whenever possible.

Definition 1.1. Assume that X is a set and $d : X \times X \to [0, \infty)$ is a function that satisfies

(*i*) $d(x, y) = 0$ if and only if $x = y$

(*ii*) $d(x, y) = d(y, x)$

(*iii*) $d(x, y) \leq d(x, z) + d(z, y)$

for all $x, y, z \in X$. Then the function d is said to be a *metric* on X, and (X, d) is called a *metric space*. When it is clear what d is, we often simply write X instead of (X, d). Let Y be a subset of X and let d^* be the restriction of d to $Y \times Y$. Then (Y, d^*) is a metric space and it is called a *subspace* of (X, d). In this case it is customary to write (Y, d) instead of (Y, d^*).

S.-M. Jung, *Ulam's Conjecture on Invariance of Measure in the Hilbert Cube*, Frontiers in Mathematics, https://doi.org/10.1007/978-3-031-30886-4_1

For each positive integer n, let $\mathbb{R}^n$ be the n-dimensional space of all ordered n-tuples of real numbers. We define the function $d_e : \mathbb{R}^n \times \mathbb{R}^n \to [0, \infty)$ by

$$d_e(x, y) = \left(\sum_{i=1}^{n} (x_i - y_i)^2 \right)^{1/2}$$

for all $x = (x_1, x_2, \ldots, x_n), y = (y_1, y_2, \ldots, y_n) \in \mathbb{R}^n$. Then d_e satisfies all the conditions in Definition 1.1, i.e., it is a metric on $\mathbb{R}^n$ which is called the *Euclidean distance* for $\mathbb{R}^n$.

Any set X can be considered a metric space in a simple way. For example, we define the function $d : X \times X \to [0, \infty)$ by

$$d(x, y) = \begin{cases} 0 & (\text{for } x = y), \\ 1 & (\text{for } x \neq y) \end{cases}$$

for all $x, y \in X$. It is easy to check that d is a metric on the set X. This metric is called the *discrete metric.*

Let x be a fixed element of a metric space (X, d) and let r be a positive real number. Then we define the *open ball*, with center x and radius r, by $B_r(x) = \{y \in X : d(y, x) < r\}$. Similarly, we define the *closed ball*, with center x and radius r, by $\overline{B}_r(x) = \{y \in X : d(y, x) \leq r\}$.

Definition 1.2. Assume that K and U are subsets of a metric space (X, d).

(i) U is said to be *open* if and only if for every $x \in U$, there exists a real number $\varepsilon > 0$ such that $B_\varepsilon(x) \subset U$;

(ii) K is said to be *closed* if and only if its complement $X \setminus K$ is open.

We will often write "U is open in X" and "K is closed in X" instead of "U is an open subset of X" and "K is a closed subset of X," respectively.

Lemma 1.3. *Assume that x is an element of a metric space (X, d) and r is a positive real number.*

(i) *The open ball $B_r(x)$ is open in X.*

(ii) *The closed ball $\overline{B}_r(x)$ is closed in X.*

Proof. (i) For any element $y \in B_r(x)$, we see that $d(y, x) < r$ and we set $\varepsilon = r - d(y, x) > 0$. Assume that z is an arbitrary element of $B_\varepsilon(y)$. Then we have

$$d(z, x) \leq d(z, y) + d(y, x) < r - d(y, x) + d(y, x) = r,$$

which implies that $z \in B_r(x)$, i.e., $y \in B_\varepsilon(y) \subset B_r(x)$. Thus, according to Definition 1.2 (i), $B_r(x)$ is an open subset of X.

We encourage the reader to do his own proof of (ii). □

Definition 1.4. Assume that X is a set and $\mathcal{T}$ is a collection of subsets of X. The collection $\mathcal{T}$ is said to be a *topology* for X if $\mathcal{T}$ satisfies the following axioms:

(i) $\emptyset \in \mathcal{T}$ and $X \in \mathcal{T}$.

(ii) If $U_1 \in \mathcal{T}$ and $U_2 \in \mathcal{T}$, then $U_1 \cap U_2 \in \mathcal{T}$.

(iii) If $\mathcal{U}$ is an arbitrary sub-collection of $\mathcal{T}$, then $\bigcup_{U \in \mathcal{U}} U \in \mathcal{T}$.

Then $(X, \mathcal{T})$ is called a *topological space*. When it is clear what $\mathcal{T}$ is, we often simply write X instead of $(X, \mathcal{T})$. Let Y be a subset of X and let $\mathcal{T}_Y$ be the collection $\{U \cap Y : U \in \mathcal{T}\}$. Then $\mathcal{T}_Y$ is a topology for Y and it is said to be the *relative topology* for Y induced by $\mathcal{T}$. Moreover, $(Y, \mathcal{T}_Y)$ is said to be a *subspace* of $(X, \mathcal{T})$. In particular, a subset U of X is called *open* in X if and only if $U \in \mathcal{T}$, and a subset K of X is called *closed* in X if and only if $X \setminus K \in \mathcal{T}$.

Let $\mathcal{P}(X)$ be the power set of a set X, i.e., the collection of all subsets of X. Then $\mathcal{P}(X)$ is a topology for X and it is called the *discrete topology* for X. Let $\mathcal{T} = \{\emptyset, X\}$. Then $\mathcal{T}$ is also a topology for X and it is called the *trivial topology* or the *indiscrete topology* for X. In general, any set can have at least two obvious topologies: the discrete topology and the indiscrete topology.

Let (X, d) be a metric space and let $\mathcal{T}(d)$ be the collection of all open subsets of X defined by Definition 1.2. It is then easy to prove that $(X, \mathcal{T}(d))$ is a topological space. So $\mathcal{T}(d)$ is said to be the topology for X generated by d.

Definition 1.5.

(i) Let (X, d) be a metric space and let $\mathcal{T}(d)$ be the collection of all open subsets of X defined by Definition 1.2. The collection $\mathcal{T}(d)$ is a topology for X and it is said to be the *topology for X generated by d*.

(ii) A topological space $(X, \mathcal{T})$ is said to be *metrizable* if there exists a metric d on X whose collection $\mathcal{T}(d)$ of generated open sets is exactly the given topology $\mathcal{T}$.

For this reason, we often treat the metric space (X, d) as a topological space $(X, \mathcal{T}(d))$. We note that the discrete topology is generated by the discrete metric but the indiscrete topology cannot be generated by any metric.

Definition 1.6. Let K and U be subsets of a topological space X.

(i) A point $x \in X$ is called an *interior point* of U if there exists an open subset V of X such that $x \in V \subset U$.

(ii) A point $x \in X$ is called a *limit point* of K if every open set containing x intersects K in a point distinct from x.

(iii) The set of all interior points of U is called the *interior* of U and is denoted by U°.

(iv) The union of K and the set of all its limit points is called the *closure* of K and is denoted by $\overline{K}$.

Theorem 1.7. *Let S be a subset of a topological space X.*

(i) *The interior of S is the union of all open subsets of X that are included in S, i.e.,*

$$S^\circ = \bigcup \{U : U \text{ is an open subset of } X \text{ included in } S\}.$$

(ii) *The closure of S is the intersection of all closed subsets of X that include S, i.e.,*

$$\overline{S} = \bigcap \{K : K \text{ is a closed subset of } X \text{ including } S\}.$$

Let $\mathbb{N}$ be the set of all positive integers and let $\mathbb{N}_0 = \mathbb{N} \cup \{0\}$. A sequence $\{x_i\}_{i\in\mathbb{N}}$ of elements of a subset S of a topological space X is said to *converge* to an element x of X if for every open subset U of X containing x, there exists a positive integer N such that $x_i \in U \cap S$ for all $i \geq N$.

The proof of the following theorem is left as an exercise for the reader.

Theorem 1.8. *The closure of a subset S of a topological space is the set of limits of all convergent sequences of elements of S.*

Definition 1.9. Let X be a metric space.

(i) A subset D of X is called a *dense* subset of X (or *dense* in X) if $\overline{D} = X$.

(ii) X is called *separable* if there exists a countable subset of X that is dense in X.

The following characterization of denseness is easy to prove and its proof will be left as an exercise.

Theorem 1.10. *Assume that D is a subset of a metric space X. Then D is dense in X if and only if each nonempty open subset of X intersects D.*

The following theorem gives an important property corresponding to separability for metric spaces.

Theorem 1.11. *A metric space X is separable if and only if there exists a countable collection $\mathcal{U}$ of open subsets of X such that each open subset of X can be expressed as the union of a sub-collection of $\mathcal{U}$.*

Proof. Assume that the metric space (X, d) is separable. Then there exists a countable dense subset $\{c_i : i \in \mathbb{N}\}$ of X. Let $\{r_i : i \in \mathbb{N}\}$ be the set of all positive rational numbers. Moreover, we define a countable collection $\mathcal{U}$ of open subsets of X by $\mathcal{U} = \{B_{r_j}(c_i) : i, j \in \mathbb{N}\}$.

We now assert that each open subset of X can be expressed as the union of a sub-collection of $\mathcal{U}$. Indeed, we will prove that if V is an open subset of X and $x \in V$, then there exists a $U \in \mathcal{U}$ such that $x \in U \subset V$. Assume that V is an arbitrary open subset of X and $x \in V$. Then there exists an $\varepsilon > 0$ such that $B_\varepsilon(x) \subset V$. Since $\{c_i : i \in \mathbb{N}\}$ is dense in X, there exists a c_i that satisfies $d(c_i, x) < \frac{1}{3}\varepsilon$. Now we note that $B_{(2/3)\varepsilon}(c_i) \subset V$. Let r_j be a positive rational number satisfying $\frac{1}{3}\varepsilon < r_j < \frac{2}{3}\varepsilon$. Then we see that $x \in B_{(1/3)\varepsilon}(c_i) \subset B_{r_j}(c_i) \subset B_{(2/3)\varepsilon}(c_i) \subset V$, i.e., if we set $U = B_{r_j}(c_i)$, then $U \in \mathcal{U}$ and $x \in U \subset V$.

Now, we assume that there exists a countable collection $\mathcal{U} = \{U_i : i \in \mathbb{N}\}$ of open subsets of X such that each open subset of X can be expressed as the union of a sub-collection of $\mathcal{U}$, where each U_i is nonempty. For every $i \in \mathbb{N}$, we choose a $c_i \in U_i$ and define $D = \{c_i : i \in \mathbb{N}\}$. Assume that V is an arbitrary nonempty open subset of X and $x \in V$. Then there exists a U_i such that $x \in U_i \subset V$. Moreover, $c_i \in V$. Hence, $V \cap D \neq \emptyset$. Thus, due to Theorem 1.10, D is a countable dense subset of X. Therefore, the metric space X is separable. □

Now we will introduce the concept of covering of a set, which is a terminology commonly used in topology. If S is a subset of a topological space $(X, \mathcal{T})$, then a *covering* of S is a collection $\{U_\lambda : \lambda \in \Lambda\}$ of subsets of X satisfying $S \subset \bigcup_{\lambda \in \Lambda} U_\lambda$, where Λ is some index set. In this case, we say that the sets U_λ's cover S. A covering $\{U_\lambda : \lambda \in \Lambda\}$ of S is called an *open covering* if each of its members is an open set, i.e., $U_\lambda \in \mathcal{T}$ for all $\lambda \in \Lambda$.

Let $\mathcal{U}$ be a covering of S. If $\mathcal{V}$ is a sub-collection of $\mathcal{U}$ and $S \subset \bigcup_{V \in \mathcal{V}} V$, then $\mathcal{V}$ is sometimes referred to as a *sub-covering* of S.

Theorem 1.12 (Lindelöf). *Let X be a separable metric space. If $\mathcal{U}$ is an open covering of a subset S of X, then there exists a countable sub-collection of $\mathcal{U}$ that also covers S.*

Proof. Assume that $\mathcal{U}$ is an open covering of S. Due to Theorem 1.11, there exists a countable collection $\mathcal{V} = \{V_i : i \in \mathbb{N}\}$ of open subsets of X such that each open subset of X is the union of a sub-collection of $\mathcal{V}$.

Let $J = \{i \in \mathbb{N} : V_i \subset U$ for at least one $U \in \mathcal{U}\}$. We will prove that $S \subset \bigcup_{i \in J} V_i$. For any $x \in S$, there exists a $U_x \in \mathcal{U}$ that includes x. Since the open set U_x is the union of a sub-collection of $\mathcal{V}$, we can choose a $k \in \mathbb{N}$ such that $x \in V_k \subset U_x$. Hence, $k \in J$ and $x \in \bigcup_{i \in J} V_i$. For any $j \in J$, we choose an open set $U_j \in \mathcal{U}$ with $V_j \subset U_j$. Now it holds that $S \subset \bigcup_{i \in J} V_i \subset \bigcup_{i \in J} U_i$. Obviously, $\{U_i : i \in J\}$ is a countable sub-collection of $\mathcal{U}$ that covers S. □

1.2 Compactness for Metric Spaces

The Heine–Borel theorem states that each subset of Euclidean space $\mathbb{R}$ is compact if and only if it is closed and bounded. In general topology, compactness is a concept that generalizes the concept of closed and bounded subsets of the Euclidean space $\mathbb{R}$ to topological spaces.

A sequence $\{x_i\}_{i \in \mathbb{N}}$ in a topological space X is said to *converge* to an element x of X if for every open subset U of X containing x, there exists a positive integer N such that $x_i \in U$ for all integers $i \geq N$.

Definition 1.13.

(i) A topological space X is called *compact* if every open covering of X contains a finite sub-collection that covers X.

(ii) A topological space X is called *sequentially compact* if every sequence in X has a subsequence that converges to an element of X.

(iii) A topological space X is said to have the *Bolzano–Weierstrass property* if every infinite subset of X has a limit point in X.

Now we will prove in the following series of theorems that the properties (i), (ii), and (iii) are equivalent to one another for metric spaces.

The symbol $\{x_i\}_{i \in \mathbb{N}}$ is used to denote a sequence whose terms are x_i's, while the symbol $\{x_i : i \in \mathbb{N}\}$ is used to denote a set whose elements are x_i's. The use of these symbols will not cause any misunderstanding.

Theorem 1.14. *A metric space is sequentially compact if and only if it has the Bolzano–Weierstrass property.*

Proof. Let X be a metric space that is sequentially compact. We assert that each infinite subset A of X has a limit point in X. Since A is infinite, we can choose a sequence $\{x_i\}_{i \in \mathbb{N}}$ of distinct elements of A. Due to the sequential compactness of X, the sequence $\{x_i\}_{i \in \mathbb{N}}$ has a subsequence which converges to an element x of

X. Then x is a limit point of the set of the subsequence, which is a subset of A, i.e., x is a limit point of A. Hence, X has the Bolzano–Weierstrass property.

We now assume that the metric space X has the Bolzano–Weierstrass property. Let $\{x_i\}_{i\in\mathbb{N}}$ be an arbitrary sequence in X. If $\{x_i\}_{i\in\mathbb{N}}$ has a term that repeats infinitely, then it has a constant subsequence, and this subsequence is obviously convergent. If no term of $\{x_i\}_{i\in\mathbb{N}}$ repeats infinitely, then the set A of terms in this sequence is infinite. Since X has the Bolzano–Weierstrass property, the set A has a limit point x and it is not difficult to choose a subsequence from $\{x_i\}_{i\in\mathbb{N}}$ that converges to x. □

Theorem 1.15. *Every compact metric space has the Bolzano–Weierstrass property.*

Proof. Let A be an arbitrary infinite subset of a compact metric space X. We assume that A has no limit point. Then every element of X is not a limit point of A, so every element of X is the center of an open ball that contains no element of A different from that center. The collection of all these open balls is an open covering of X, and by the compactness of X, there exists a finite sub-collection that covers X. Since A is contained in the set of all centers of open balls in this sub-collection, A is obviously finite. This contradicts the assumption that A is infinite. Therefore, we may conclude that A has a limit point, i.e., X has the Bolzano–Weierstrass property. □

Definition 1.16. A metric space (X, d) is called *totally bounded* if it has the following property: For any $\varepsilon > 0$, there exists a finite set F_ε contained in X such that

$$X = \bigcup_{x\in F_\varepsilon} B_\varepsilon(x),$$

where $B_\varepsilon(x)$ denotes the open ball with center x and radius ε. In this case, F_ε is called an *ε-net*.

Theorem 1.17. *Every sequentially compact metric space is totally bounded.*

Proof. Let X be a sequentially compact metric space. Given $\varepsilon > 0$, choose an element x_1 from X and construct the open ball $B_\varepsilon(x_1)$. If this open ball contains all elements of X, then the single-element set $\{x_1\}$ is an ε-net. If there are elements outside of $B_\varepsilon(x_1)$, then let x_2 be one of those elements outside of $B_\varepsilon(x_1)$ and form the set $B_\varepsilon(x_1) \cup B_\varepsilon(x_2)$. If this union contains all elements of X, then the two-element set $\{x_1, x_2\}$ is an ε-net.

If we could continue like this infinitely, then the sequence $\{x_1, x_2, \ldots\}$ would be a sequence without a convergent subsequence, contrary to the assumed sequential compactness of X. Therefore, a union of the form $B_\varepsilon(x_1) \cup B_\varepsilon(x_2) \cup \cdots \cup$

$B_\varepsilon(x_n)$ necessarily contains all elements of X. Then the finite set $\{x_1, x_2, \ldots, x_n\}$ is an ε-net, so X is totally bounded. □

Let (X, d) be a metric space. A subset B of X is said to be *bounded* if there exists a positive number M such that $d(x, y) \le M$ for all $x, y \in B$. The *diameter* of B is defined as

$$d(B) = \begin{cases} 0 & (\text{for } B = \emptyset), \\ \sup\{d(x,y) : x, y \in B\} & (\text{for a nonempty bounded set } B), \\ \infty & (\text{for an unbounded set } B). \end{cases}$$

Let $\{U_\lambda : \lambda \in \Lambda\}$ be an open covering of a metric space X. A positive real number a is called a *Lebesgue number* for the open covering $\{U_\lambda : \lambda \in \Lambda\}$ if every subset of X whose diameter is smaller than a is contained in at least one U_λ.

The following lemma is known as Lebesgue's covering lemma.

Lemma 1.18 (Lebesgue's Covering Lemma). *Let X be a sequentially compact metric space. Then every open covering of X has a Lebesgue number.*

Proof. Let X be a sequentially compact metric space and let $\{U_\lambda : \lambda \in \Lambda\}$ be an open covering of X. A subset of X is said to be *big* if it is not contained in any U_λ. When there are no big sets, any positive real number serves as our Lebesgue number a. We may thus assume that there are big sets, and we define a' as the greatest lower bound of their diameters. It is obvious that $0 \le a' \le \infty$.

We will prove that $a' > 0$. If $a' = \infty$, then any positive real number can be a, and if a' is a positive real number, then we can take $a = \frac{1}{2}a'$. Therefore, we assume that $a' = 0$ and derive a contradiction from this assumption. Since every big set has at least two elements, we conclude from $a' = 0$ that for every positive integer n there is a big set B_n such that $0 < d(B_n) < \frac{1}{n}$, where $d(B_n)$ denotes the diameter of B_n.

We now choose an element x_n of each B_n. Since X is sequentially compact, the sequence $\{x_n\}_{n \in \mathbb{N}}$ has a subsequence which converges to some element x of X. Then the element x belongs to at least one set U_{λ_0} in the open covering, and since U_{λ_0} is open, x is the center of some open ball $B_r(x)$ contained in U_{λ_0}. Let $B_{(1/2)r}(x)$ be the concentric open ball with radius $\frac{1}{2}r$. Since the subsequence of $\{x_n\}_{n \in \mathbb{N}}$ converges to x, there are infinitely many positive integers n for which x_n belongs to $B_{(1/2)r}(x)$. Let n_0 be one of these positive integers, which is so large that $\frac{1}{n_0} < \frac{1}{2}r$. Since $d(B_{n_0}) < \frac{1}{n_0} < \frac{1}{2}r$, we see that $B_{n_0} \subset B_r(x) \subset U_{\lambda_0}$. This contradicts the fact that B_{n_0} is a big set, and we complete the proof. □

Finally we will prove that sequential compactness implies compactness.

Theorem 1.19. *Every sequentially compact metric space is compact.*

Proof. Let X be a sequentially compact metric space and let $\{U_\lambda : \lambda \in \Lambda\}$ be an open covering of X. According to Lemma 1.18, this open covering has a Lebesgue number $a > 0$. We put $\varepsilon = \frac{1}{3}a$ and use Theorem 1.17 to find an ε-net

$$F_\varepsilon = \{x_1, x_2, \ldots, x_n\}.$$

For any $k \in \{1, 2, \ldots, n\}$, we have $d(B_\varepsilon(x_k)) \leq 2\varepsilon = \frac{2}{3}a < a$. By the definition of Lebesgue number, for every k, we may select a U_{λ_k} such that $B_\varepsilon(x_k) \subset U_{\lambda_k}$. Since every element of X belongs to one of the $B_\varepsilon(x_k)$'s, the collection $\{U_{\lambda_1}, U_{\lambda_2}, \ldots, U_{\lambda_n}\}$ is a finite sub-covering of X. Therefore, X is compact. □

In this way we proved that compactness, sequential compactness, and the Bolzano–Weierstrass property are equivalent for any metric space.

1.3 Compact Topological Spaces

Let X be a topological space. We recall that a collection $\{U_\lambda : \lambda \in \Lambda\}$ of open subsets of X is called an open covering of X if every element of X belongs to at least one U_λ. A sub-collection of an open covering that is itself an open covering is called an open sub-covering.

Theorem 1.20. *Let C be a compact subset of a topological space X. If C is included in a subset K of X, then C is a compact subset of K.*

Proof. Assume that $\mathcal{V}$ is an arbitrary covering of C and each member of $\mathcal{V}$ is open in K, where K is equipped with the relative topology. Then there exists an open subset U of X such that $V = U \cap K$ for each $V \in \mathcal{V}$. Let $\mathcal{U}$ be the collection of these open sets U's. Then $\mathcal{U}$ is an open covering of C. Since C is a compact subset of X, there exists a finite sub-collection $\{U_1, U_2, \ldots, U_n\}$ that covers C. Thus, $\{V_1, V_2, \ldots, V_n\}$ is a finite sub-collection of $\mathcal{V}$ that obviously covers C, where $V_i = U_i \cap K$ for all $i \in \{1, 2, \ldots, n\}$. Therefore, C is a compact subset of K. □

Indeed, the reverse assertion of Theorem 1.20 also holds. More precisely, if C is a compact subset of K, where K is a subset of X equipped with the relative topology, then C is a compact subset of X.

We note that each closed subset of a compact metric space is compact. In the following theorem, we will prove that this is generally true in all topological spaces.

Theorem 1.21. *Every closed subset K of any compact topological space X is compact.*

Proof. Let $\mathcal{U} = \{U_\lambda : \lambda \in \Lambda\}$ be an arbitrary open covering of K. Then $\mathcal{V} := \mathcal{U} \cup \{X \setminus K\}$ is an open covering of X. Since X is compact, $\mathcal{V}$ contains a finite sub-collection $\{U_{\lambda_1}, U_{\lambda_2}, \ldots, U_{\lambda_n}\} \cup \{X \setminus K\}$ that covers X, where each $\lambda_i \in \Lambda$. Hence, we see that $K \subset U_{\lambda_1} \cup U_{\lambda_2} \cup \cdots \cup U_{\lambda_n}$, which implies that $\mathcal{U}$ contains a finite sub-collection that covers K. Therefore, K is compact. □

The intuitive idea behind the notion of continuity of functions is to maintain the proximity of points. Expressing this idea as sentences in general topology, we get the following definition.

Definition 1.22. Let X and Y be topological spaces and let $f : X \to Y$ be a function.

(i) The function f is called *continuous* at $x_0 \in X$ if for each open subset V of Y with $f(x_0) \in V$, there exists an open subset U of X such that $x_0 \in U$ and $f(U) \subset V$.

(ii) If f is continuous at each element of X, then f is said to be a *continuous function.*

(iii) The function f is a continuous function if and only if the pre-image of each open subset of Y is an open subset of X.

The equivalence of Definition 1.22 (ii) and (iii) is well known and is left as an exercise for the reader.

The following theorem states that the images of compact sets under every continuous function are also compact.

Theorem 1.23. *Let X and Y be topological spaces. If $f : X \to Y$ is a continuous function and K is a compact subset of X, then $f(K)$ is a compact subset of Y.*

Proof. Let $\mathcal{V} = \{V_\lambda : \lambda \in \Lambda\}$ be an arbitrary open covering of $f(K)$, where each V_λ is an open subset of Y. Since

$$K \subset f^{-1}(f(K)) \subset f^{-1}\left(\bigcup_{\lambda \in \Lambda} V_\lambda\right) = \bigcup_{\lambda \in \Lambda} f^{-1}(V_\lambda),$$

the collection $\mathcal{U} = \{f^{-1}(V_\lambda) : \lambda \in \Lambda\}$ is an open covering of the compact set K. Hence, the compactness of K implies that the open covering $\mathcal{U}$ of K contains a finite sub-covering $\{f^{-1}(V_{\lambda_1}), f^{-1}(V_{\lambda_2}), \ldots, f^{-1}(V_{\lambda_n})\}$ of K, where each $\lambda_i \in \Lambda$, i.e.,

$$K \subset \bigcup_{i=1}^{n} f^{-1}(V_{\lambda_i}),$$

from which it follows that

$$f(K) \subset f\left(\bigcup_{i=1}^{n} f^{-1}(V_{\lambda_i})\right) = \bigcup_{i=1}^{n} f(f^{-1}(V_{\lambda_i})) \subset \bigcup_{i=1}^{n} V_{\lambda_i}.$$

Therefore, each open covering $\mathcal{V}$ of $f(K)$ contains a finite open sub-covering $\{V_{\lambda_1}, V_{\lambda_2}, \ldots, V_{\lambda_n}\}$. That is, $f(K)$ is a compact subset of Y. □

Proving that a topological space is compact by referring directly to the definition is sometimes difficult. The following theorem gives an equivalent form of compactness that is often easier to apply.

Theorem 1.24. *A topological space is compact if and only if every collection of closed sets with empty intersection has a finite sub-collection with empty intersection.*

Proof. We assume that X is a compact topological space and $\{K_\lambda : \lambda \in \Lambda\}$ is an arbitrary collection of closed sets with empty intersection. Then, we have

$$\bigcap_{\lambda \in \Lambda} K_\lambda = \emptyset.$$

Thus, we get

$$X = X \setminus \bigcap_{\lambda \in \Lambda} K_\lambda = \bigcup_{\lambda \in \Lambda} (X \setminus K_\lambda) = \bigcup_{\lambda \in \Lambda} U_\lambda,$$

where we set $U_\lambda = X \setminus K_\lambda$ for all $\lambda \in \Lambda$. Hence, $\{U_\lambda : \lambda \in \Lambda\}$ is an open covering of X. Then, by Definition 1.13 (i), the open covering $\{U_\lambda : \lambda \in \Lambda\}$ of X contains a finite sub-covering $\{U_{\lambda_1}, U_{\lambda_2}, \ldots, U_{\lambda_n}\}$, i.e.,

$$X \subset \bigcup_{i=1}^{n} U_{\lambda_i}.$$

Therefore,

$$\bigcap_{i=1}^{n} K_{\lambda_i} = \bigcap_{i=1}^{n} (X \setminus U_{\lambda_i}) = X \setminus \bigcup_{i=1}^{n} U_{\lambda_i} = \emptyset,$$

i.e., the collection $\{K_\lambda : \lambda \in \Lambda\}$ has a finite sub-collection $\{K_{\lambda_1}, K_{\lambda_2}, \ldots, K_{\lambda_n}\}$ with empty intersection.

Conversely, we assume that X is a topological space and every collection of closed sets with empty intersection has a finite sub-collection with empty intersection. Let $\{U_\lambda : \lambda \in \Lambda\}$ be an arbitrary open covering of X. If we define

$K_\lambda = X \setminus U_\lambda$ for all $\lambda \in \Lambda$, then $\{K_\lambda : \lambda \in \Lambda\}$ is a collection of closed sets with empty intersection:

$$\bigcap_{\lambda\in\Lambda} K_\lambda = \bigcap_{\lambda\in\Lambda}(X \setminus U_\lambda) = X \setminus \bigcup_{\lambda\in\Lambda} U_\lambda = \emptyset.$$

Thus, by our assumption, the collection $\{K_\lambda : \lambda \in \Lambda\}$ has a finite sub-collection $\{K_{\lambda_1}, K_{\lambda_2}, \ldots, K_{\lambda_n}\}$ with empty intersection. Hence, we have

$$X = X \setminus \bigcap_{i=1}^{n} K_{\lambda_i} = \bigcup_{i=1}^{n}(X \setminus K_{\lambda_i}) = \bigcup_{i=1}^{n} U_{\lambda_i}.$$

That is, $\{U_{\lambda_1}, U_{\lambda_2}, \ldots, U_{\lambda_n}\}$ is a finite sub-covering of X. Therefore, X is compact. □

The statement "every collection of closed sets with empty intersection has a finite sub-collection with empty intersection" is equivalent to the statement "for every collection $\{K_\lambda : \lambda \in \Lambda\}$ of closed sets, if $\bigcap_{\lambda\in\Lambda} K_\lambda = \emptyset$ then there exists a finite sub-collection $\{K_{\lambda_1}, K_{\lambda_2}, \ldots, K_{\lambda_n}\}$ such that $\bigcap_{i=1}^{n} K_{\lambda_i} = \emptyset$," which is again equivalent to the statement "for every collection $\{K_\lambda : \lambda \in \Lambda\}$ of closed sets, if $\bigcap_{i=1}^{n} K_{\lambda_i} \neq \emptyset$ for any finite sub-collection $\{K_{\lambda_1}, K_{\lambda_2}, \ldots, K_{\lambda_n}\}$ then $\bigcap_{\lambda\in\Lambda} K_\lambda \neq \emptyset$."

Indeed, a collection of subsets of a nonempty set is said to have the *finite intersection property* if every finite sub-collection has nonempty intersection. Hence, the statement "every collection of closed sets with empty intersection has a finite sub-collection with empty intersection" is equivalent to the statement "every collection of closed sets with the finite intersection property has nonempty intersection."

This concept allows us to express Theorem 1.24 as follows.

Theorem 1.25. *A topological space is compact if and only if every collection of closed sets with the finite intersection property has nonempty intersection.*

Definition 1.26. Let X be a topological space.

(i) A collection of open subsets of X is called an *open base* for X if every open set is a union of some members of this collection. A collection of closed subsets of X is called a *closed base* for X if the collection of all complements of its members is an open base.

(ii) A collection of open subsets of X is called an *open subbase* for X if its finite intersections form an open base. A collection of closed subsets of X is called a *closed subbase* for X if the collection of all complements of its members is an open subbase.

By Theorem 1.11, every separable metric space has a countable open base.

An open covering of a topological space, whose members are all in some given open base, is called a *basic open covering*. An open covering whose members are all in some open subbase is called a *subbasic open covering*.

Theorem 1.27. *Let X be a topological space.*

(i) *X is compact if and only if every basic open covering has a finite sub-covering.*

(ii) *X is compact if and only if every collection of basic closed sets with the finite intersection property has nonempty intersection.*

Proof. (i) We assume that every basic open covering of X has a finite sub-covering. Let $\{U_\lambda : \lambda \in \Lambda\}$ be an arbitrary open covering of X and let $\{V_\gamma : \gamma \in \Gamma\}$ be an open base for X. For each $\lambda \in \Lambda$, U_λ is the union of certain V_γ's, and the collection of all these V_γ's is a basic open covering of X. By our assumption, this collection of V_γ's has a finite sub-covering. For any member of this finite sub-covering, we can choose a U_λ that contains it. The resulting collection of U_λ's is obviously a finite open sub-covering of X. Therefore, we conclude that X is compact.

Now we assume that the topological space X is compact. Since each basic open covering is also an open covering of the compact topological space X, it has a finite sub-covering.

(ii) If X is compact, then it follows from Theorem 1.25 that every collection of closed sets with the finite intersection property has nonempty intersection. Let $\{B_\lambda : \lambda \in \Lambda\}$ be an arbitrary collection of basic closed sets with the finite intersection property. Since each basic closed set B_λ is a closed set, according to the last argument, the collection $\{B_\lambda : \lambda \in \Lambda\}$ has nonempty intersection.

Now we assume that every collection of basic closed sets with the finite intersection property has nonempty intersection. Let $\{B_\lambda : \lambda \in \Lambda\}$ be an arbitrary collection of basic closed sets with the finite intersection property. Then the collection $\{B_\lambda : \lambda \in \Lambda\}$ has nonempty intersection, i.e., $\bigcap_{\lambda \in \Lambda} B_\lambda \neq \emptyset$. Assume that $\{K_\gamma : \gamma \in \Gamma\}$ is an arbitrary collection of closed sets with the finite intersection property.

Then $\{X \setminus B_\lambda : \lambda \in \Lambda\}$ is an open base for X and $\{X \setminus K_\gamma : \gamma \in \Gamma\}$ is a collection of open subsets of X. For every $\gamma \in \Gamma$, there exists a nonempty index subset Λ_γ of Λ such that

$$X \setminus K_\gamma = \bigcup_{\lambda \in \Lambda_\gamma} X \setminus B_\lambda.$$

Thus, we have

$$\bigcup_{\gamma \in \Gamma} X \setminus K_\gamma \subset \bigcup_{\lambda \in \Lambda} X \setminus B_\lambda,$$

which implies that

$$\emptyset \neq \bigcap_{\lambda \in \Lambda} B_\lambda \subset \bigcap_{\gamma \in \Gamma} K_\gamma.$$

Therefore, by Theorem 1.25, X is compact. □

As the following theorem shows, we can partially extend the previous theorem to the theorem related to open subbase.

Theorem 1.28. *A topological space is compact if every subbasic open covering has a finite sub-covering, or equivalently, if every collection of subbasic closed sets with the finite intersection property has nonempty intersection.*

Proof. From Theorems 1.24 and 1.25, it is easy to see that the two sufficient conditions mentioned in this theorem for compactness are equivalent. Consider a closed subbase for the topological space, and assume that $\{B_i\}_i$ is its generated closed base, i.e., the collection of all finite unions of its members. We assume that every collection of subbasic closed sets with the finite intersection property has nonempty intersection, and we prove from this assumption that every collection of B_i's with the finite intersection property also has nonempty intersection. Due to Theorem 1.27, this is sufficient to prove our theorem.

Let $\{B_j\}_j$ be a collection of B_i's with the finite intersection property. We must show that $\bigcap_j B_j \neq \emptyset$. We use Zorn's lemma to show that $\{B_j\}_j$ is contained in some collection $\{B_k\}_k$ of B_i's which is maximal with respect to having the finite intersection property, in the sense that $\{B_k\}_k$ has this property and any collection of B_i's which properly contains $\{B_k\}_k$ fails to have this property. The argument runs as follows. Consider the family of all collections of B_i's which contain $\{B_j\}_j$ and have the finite intersection property. This is a partially ordered set with respect to collection inclusion. If we consider a chain in this partially ordered set, the union of all collections in it is a collection of B_i's which contains every member of the chain and has the finite intersection property, as we see from the fact that every finite collection of its sets is contained in some member of the chain and that the member has the finite intersection property. We conclude that every chain in our partially ordered set has an upper bound, so Zorn's lemma guarantees that the partially ordered set has a maximal element. This argument yields the existence of a collection $\{B_k\}_k$ with the properties stated above. Since $\bigcap_k B_k \subset \bigcap_j B_j$, it now suffices to show that $\bigcap_k B_k \neq \emptyset$.

Each B_k is a finite union of members of our closed subbase, for example, $B_1 = S_1 \cup S_2 \cup \cdots \cup S_n$. It now suffices to show that at least one of the members $S_1, S_2, \ldots, S_n$ belongs to the collection $\{B_k\}_k$. For if we obtain such a set for each B_k, the resulting collection of subbasic closed sets will have the finite intersection property (since it is contained in $\{B_k\}_k$), and therefore, by our hypothesis relating to the subbasic closed sets, it will have nonempty intersection; since this nonempty intersection will be a subset of $\bigcap_k B_k$, we shall know that $\bigcap_k B_k$ is itself nonempty.

We finish the proof by showing that at least one of the members $S_1, S_2, \ldots, S_n$ does in fact belong to the collection $\{B_k\}_k$. We assume that each of these sets is not in this collection, and we deduce a contradiction from this assumption. Since S_1 is a subbasic closed set, it is also a basic closed set, and since it is not in the collection $\{B_k\}_k$, the collection $\{B_k\}_k \cup \{S_1\}$ is a collection of B_i's which properly contains $\{B_k\}_k$. By the maximality property of $\{B_k\}_k$, the collection $\{B_k\}_k \cup \{S_1\}$ lacks the finite intersection property, so S_1 is disjoint from the intersection of some finite collection of B_k's. If we do this for each of the sets $S_1, S_2, \ldots, S_n$, we see that B_1—the union of these sets—is disjoint from the intersection of the total finite collection of all the B_k's which arise in this way. This contradicts the finite intersection property for the collection $\{B_k\}_k$ and completes the proof. □

As we can guess from the complexity of the proof, the scope of this theorem is very wide. We will show the practicability of this theorem by applying the theorem to provide a simple proof of the classical Heine–Borel theorem.

Theorem 1.29 (Heine–Borel). *Every subset of the Euclidean space $\mathbb{R}$ is compact if and only if it is closed and bounded.*

Proof. It is well known that if a subset of $\mathbb{R}$ is compact, then it is closed and bounded. Therefore, we will prove that a subset of $\mathbb{R}$, which is closed and bounded, is compact. We note that a closed and bounded subset of $\mathbb{R}$ is a closed subset of some closed interval $[a, b]$. Thus, we will prove that $[a, b]$ is compact under the assumption that $a < b$, without loss of generality. We note that the collection of all intervals of the form $[a, d)$ and $(c, b]$, where c and d are any real numbers such that

$a < c < b$ and $a < d < b$, is an open subbase for $[a, b]$. Therefore, the collection of all $[a, c]$'s and all $[d, b]$'s is a closed subbase for $[a, b]$.

Let $\mathcal{S} = \{[a, c_i], [d_j, b] : i, j \in \mathbb{N}\}$ be a collection of these subbasic closed sets with the finite intersection property. In view of Theorem 1.28, we will prove that the intersection of all members of $\mathcal{S}$ is nonempty. Without loss of generality, we assume that $\mathcal{S} \neq \emptyset$. If $\mathcal{S}$ contains only intervals of the form $[a, c_i]$ or only intervals of the form $[d_j, b]$, then the intersection clearly contains a or b. Hence, we assume that $\mathcal{S}$ contains intervals of both forms. Now we set $d = \sup\{d_j : j \in \mathbb{N}\}$ and we complete the proof by showing that $d \leq c_i$ for all i. If we assume that $c_{i_0} < d$ for some i_0, it then follows from the definition of d that there exists a d_{j_0} such that $c_{i_0} < d_{j_0}$. Since $[a, c_{i_0}] \cap [d_{j_0}, b] = \emptyset$, it contradicts the finite intersection property for $\mathcal{S}$ and completes the proof. □

In general, the Heine–Borel theorem holds in the Euclidean space $\mathbb{R}^n$.

Theorem 1.30. *Let n be a positive integer. Every subset of the Euclidean space $\mathbb{R}^n$ is compact if and only if it is closed and bounded.*

The Heine–Borel theorem holds for all finite-dimensional normed spaces, but in general this is not the case. It is not difficult to find metric spaces or topological spaces for which the Heine–Borel theorem does not hold. In many metric spaces, such as incomplete metric spaces, the Heine–Borel theorem does not hold. Even in complete metric spaces, the Heine–Borel theorem may not hold. For example, the Heine–Borel theorem does not hold for any infinite-dimensional Banach space. Even more surprising is that the Heine–Borel theorem may not hold for the real line unless given the usual metric.

1.4 Product of Topological Spaces

One of the ways to create a new topological space from existing topological spaces is to multiply the given spaces.

Assume that $\{X_1, X_2, \dots, X_n\}$ is a collection of sets, where n is a positive integer. The *Cartesian product* $\prod_{i=1}^{n} X_i$ is defined as the set of all finite sequences $(x_1, x_2, \dots, x_n)$, where $x_i \in X_i$ for all $i \in \{1, 2, \dots, n\}$. We note that each $x \in \prod_{i=1}^{n} X_i$ is a function defined on $\{1, 2, \dots, n\}$. We can generalize the concept of Cartesian products to any indexed collections of sets as follows:

Definition 1.31. Assume that $\{X_\lambda : \lambda \in \Lambda\}$ is an indexed collection of sets. The *Cartesian product* $\prod_{\lambda \in \Lambda} X_\lambda$ of the collection $\{X_\lambda : \lambda \in \Lambda\}$ is defined as the collection of all functions x defined on Λ such that $x(\lambda) \in X_\lambda$. For any $\lambda \in \Lambda$, X_λ is called the *λ-coordinate space*, and for any $x \in \prod_{\lambda \in \Lambda} X_\lambda$, $x(\lambda)$, also written as x_λ, is called the *λ-coordinate* of x.

We note that if $\Lambda \neq \emptyset$ and each λ-coordinate space X_λ is nonempty, then $\prod_{\lambda \in \Lambda} X_\lambda$ is nonempty by the axiom of choice.

Definition 1.32. Let $\prod_{\lambda \in \Lambda} X_\lambda$ be the Cartesian product of the collection $\{X_\lambda : \lambda \in \Lambda\}$. For any $\mu \in \Lambda$, let $\pi_\mu : \prod_{\lambda \in \Lambda} X_\lambda \to X_\mu$ be defined by $\pi_\mu(x) = x_\mu$, where x_μ denotes the μ-coordinate of x. The function π_μ is called the *projection function* of $\prod_{\lambda \in \Lambda} X_\lambda$ into the μ-coordinate space X_μ.

Assume that $\{(X_\lambda, \mathcal{T}_\lambda) : \lambda \in \Lambda\}$ is an indexed collection of topological spaces. In the following definition, we define a topology for the Cartesian product $\prod_{\lambda \in \Lambda} X_\lambda$.

Definition 1.33. Assume that $\{(X_\lambda, \mathcal{T}_\lambda) : \lambda \in \Lambda\}$ is a collection of topological spaces. We define the collection $\mathcal{S}$ of subsets of $\prod_{\lambda \in \Lambda} X_\lambda$ by

$$\mathcal{S} = \left\{\pi_\lambda^{-1}(U_\lambda) : \lambda \in \Lambda \text{ and } U_\lambda \in \mathcal{T}_\lambda\right\}.$$

The topology $\mathcal{T}$ for $X = \prod_{\lambda \in \Lambda} X_\lambda$ that has $\mathcal{S}$ as a subbase is called the *product topology* and the topological space $(X, \mathcal{T})$ is called the *product space*.

A base $\mathcal{B}$ for the product topology given in Definition 1.33 is the collection of all sets of the form

$$\bigcap_{\lambda \in \Delta} \pi_\lambda^{-1}(U_\lambda),$$

where Δ is a finite subset of the indexing set Λ, and for each $\lambda \in \Delta$, U_λ is open in X_λ. Moreover,

$$\bigcap_{\lambda \in \Delta} \pi_\lambda^{-1}(U_\lambda) = \prod_{\lambda \in \Lambda} V_\lambda, \quad \text{where} \quad V_\lambda = \begin{cases} U_\lambda & (\text{for } \lambda \in \Delta), \\ X_\lambda & (\text{for } \lambda \in \Lambda \setminus \Delta). \end{cases}$$

We note that for each nonempty open subset U of $\prod_{\lambda \in \Lambda} X_\lambda$, $\pi_\mu(U) = X_\mu$ for all $\mu \in \Lambda \setminus \Delta$.

Theorem 1.34. *Let* $\prod_{\lambda\in\Lambda} X_\lambda$ *be a nonempty product space. For each* $\mu \in \Lambda$*, the projection function*

$$\pi_\mu : \prod_{\lambda\in\Lambda} X_\lambda \to X_\mu$$

is an open continuous surjection.

Proof. By Definition 1.32, it is obvious that the projection function π_μ is a surjection. We know that if U_μ is an open subset of X_μ, then $\pi_\mu^{-1}(U_\mu)$ is a member of the defining open subbase for the product topology. It follows from this fact and Definition 1.26 that π_μ is continuous.

Assume that Δ is a finite subset of Λ and U_λ is an open subset of X_λ for all $\lambda \in \Delta$. Then, by the remark above, the set $V = \bigcap_{\lambda\in\Delta} \pi_\lambda^{-1}(U_\lambda)$ is a member of a base $\mathcal{B}$ for the product topology. Then $\pi_\mu(V) = U_\mu$ for $\mu \in \Delta$ or $\pi_\mu(V) = X_\mu$ for $\mu \in \Lambda \setminus \Delta$. In either case, $\pi_\mu(V)$ is open in X_μ. Hence, π_μ is an open function. □

We note that every coordinate space inherits many properties from the product space: every property that is invariant under continuous open surjections is inherited by every coordinate space from the product space.

A topological space X is called a *Hausdorff space* or a T_2*-space* if for any distinct elements $x, y \in X$, there exist disjoint open subsets U and V of X such that $x \in U$ and $y \in V$.

Theorem 1.35. *Let* $\prod_{\lambda\in\Lambda} X_\lambda$ *be a nonempty product space. Then* $\prod_{\lambda\in\Lambda} X_\lambda$ *is a Hausdorff space if and only if each coordinate space* X_λ *is a Hausdorff space.*

Proof. If the product space $\prod_{\lambda\in\Lambda} X_\lambda$ is a Hausdorff space, then every coordinate space X_λ is a Hausdorff space according to Theorem 1.34 and the above remark.

We now assume that the coordinate space X_λ is a Hausdorff space for all $\lambda \in \Lambda$. Assume that x and y are distinct elements of the product space $\prod_{\lambda\in\Lambda} X_\lambda$. Then $x_{\lambda_0} \neq y_{\lambda_0}$ for some $\lambda_0 \in \Lambda$. Since X_{λ_0} is a Hausdorff space, there exists a pair of disjoint open subsets U_{λ_0} and V_{λ_0} of X_{λ_0} such that $x_{\lambda_0} \in U_{\lambda_0}$ and $y_{\lambda_0} \in V_{\lambda_0}$. Then $\pi_{\lambda_0}^{-1}(U_{\lambda_0})$ and $\pi_{\lambda_0}^{-1}(V_{\lambda_0})$ are disjoint open subsets such that $x \in \pi_{\lambda_0}^{-1}(U_{\lambda_0})$ and $y \in \pi_{\lambda_0}^{-1}(V_{\lambda_0})$, which implies that the product space is a Hausdorff space. □

Tychonoff's theorem, often considered one of the most important results of general topology, states that the product of any collection of compact topological spaces is compact with respect to the product topology.

Now we have everything we need to prove Tychonoff's theorem.

Theorem 1.36 (Tychonoff). *The product space of any nonempty collection of compact topological spaces is compact with respect to the product topology.*

Proof. Assume that $\{X_\lambda : \lambda \in \Lambda\}$ is a nonempty collection of compact topological spaces and $X = \prod_{\lambda\in\Lambda} X_\lambda$ is its product space. Let $\{F_\mu\}_\mu$ be a nonempty collection of the defining closed subbase for the product topology for X. Then each F_μ is of the form $F_\mu = \prod_\lambda F_{\lambda\mu}$, where $F_{\mu\mu}$ is a closed subset of X_μ but $F_{\lambda\mu} = X_\lambda$ for all $\lambda \neq \mu$.

We assume that the collection $\{F_\mu\}_\mu$ has the finite intersection property, and by Theorem 1.28 we complete the proof by verifying that $\bigcap_\mu F_\mu \neq \emptyset$.

For a fixed $\lambda \in \Lambda$, $\{F_{\lambda\mu}\}_\mu$ is a collection of closed subsets of X_λ with the finite intersection property. Since X_λ is assumed to be compact, according to Theorem 1.25, there exists an $x_\lambda \in X_\lambda$ which belongs to $\bigcap_\mu F_{\lambda\mu}$. By applying this process to each λ, we obtain an element $x = (x_\lambda)_{\lambda\in\Lambda}$ of $\bigcap_\mu F_\mu$. □

Indeed, Tychonoff's theorem is more widely known in its generalized form:

Theorem 1.37. *Let $X = \prod_{\lambda\in\Lambda} X_\lambda$ be the nonempty product space of a nonempty collection $\{X_\lambda : \lambda \in \Lambda\}$ of topological spaces. Then X is compact if and only if each coordinate space X_λ is compact.*

1.5 Completeness

From an intuitive point of view, a space is complete when there are no "missing points." For example, the set of all rational numbers is not complete because, e.g., $\sqrt{2}$ is "missing" in it, although we can construct a Cauchy sequence of rational numbers that converges to it.

A metric space X is called complete if every Cauchy sequence in X has a limit that is also in X.

Definition 1.38. Assume that $\{c_i\}_{i\in\mathbb{N}}$ is a sequence in a metric space (X, d). Then $\{c_i\}_{i\in\mathbb{N}}$ is said to be a *Cauchy sequence* in (X, d) if $\{c_i\}_{i\in\mathbb{N}}$ satisfies the condition: for every $\varepsilon > 0$, there exists an integer $N_\varepsilon > 0$ such that if $m \geq N_\varepsilon$ and $n \geq N_\varepsilon$, then $d(c_m, c_n) < \varepsilon$.

According to Cauchy criterion for convergence in $\mathbb{R}^n$, every sequence in $\mathbb{R}^n$ converges if and only if it is a Cauchy sequence. Hence, we know that the collection of Cauchy sequences in $\mathbb{R}^n$ and the collection of convergent sequences in $\mathbb{R}^n$ are the same.

For all metric spaces, every convergent sequence is a Cauchy sequence. However, it is not true that all Cauchy sequences in every metric space necessarily converge. Since it is an important property that all Cauchy sequences converge, it is necessary to classify metric spaces with this property separately.

Definition 1.39. A metric space (X, d) is said to be *complete* if every Cauchy sequence in (X, d) converges.

The metric space $\mathbb{R}$ of real numbers with the metric given by the absolute value is complete, as is the n-dimensional Euclidean space $\mathbb{R}^n$ with the Euclidean distance. On the other hand, the metric space $\mathbb{Q}$ of rational numbers is not complete because, e.g., $\sqrt{2}$ is missing from it, although we can construct a Cauchy sequence of rational numbers that converges to it.

Theorem 1.40. *A metric space is compact if and only if it is complete and totally bounded.*

Proof. Assume that X is complete and totally bounded metric space. In view of Theorem 1.19, we will prove that X is compact by showing that every sequence has a convergent subsequence. Since X is complete, it suffices to show that every sequence in X has a Cauchy subsequence. Consider an arbitrary sequence

$$S_1 = \{x_{11}, x_{12}, x_{13}, \dots\}$$

in X. Since X is assumed to be totally bounded, there exists a finite collection of open balls, each of radius $\frac{1}{2}$, whose union equals X. Thus, we note that S_1 has a subsequence

$$S_2 = \{x_{21}, x_{22}, x_{23}, \dots\}$$

whose terms are all lie in an open ball of radius $\frac{1}{2}$. Similarly, another application of the total boundedness of X shows that S_2 has a subsequence

$$S_3 = \{x_{31}, x_{32}, x_{33}, \dots\}$$

whose terms all lie in an open ball of radius $\frac{1}{3}$. In this way we continue to form consecutive subsequences and define

$$S = \{x_{11}, x_{22}, x_{33}, \dots\}.$$

By this construction, S is clearly a Cauchy subsequence of S_1. Hence, X is a sequentially compact metric space. According to Theorem 1.19, X is a compact metric space.

Let X be a compact metric space. In view of Theorems 1.14, 1.15, and 1.17, X is sequentially compact and totally bounded. Let $S_1 = \{x_{11}, x_{12}, x_{13}, \ldots\}$ be an arbitrary Cauchy sequence in X. Since X is sequentially compact, S_1 has a convergent subsequence. That is, the Cauchy sequence S_1 is convergent in X, which implies that X is complete. $\square$

We note that a closed subspace of a complete metric space is complete. If (X, d) is a metric space and Y is a subset of X that is not closed, then there exists a sequence $\{y_i\}_{i\in\mathbb{N}}$ in Y that converges to a point $x_0 \in X \setminus Y$. Thus, $\{y_i\}_{i\in\mathbb{N}}$ is a Cauchy sequence in Y that does not converge in (Y, d). Hence, (Y, d) is not complete. These remarks are summarized in the following theorem.

Theorem 1.41. *If K is a complete subset of a metric space X, then K is closed. Moreover, if X is a complete metric space and K is a closed subset of X, then K is complete.*

1.6 Separation Properties

Let $\mathcal{F}$ be the collection of all topologies for a nonempty set X. We may introduce the set inclusion "$\subset$" for the collection $\mathcal{F}$, which is a partial order for $\mathcal{F}$, and the pair $(\mathcal{F}, \subset)$ is a partially ordered set. Then the discrete topology for X is the maximum element and the trivial topology is the minimum element of $\mathcal{F}$.

In the following definition, we define a finite sub-collection $\mathcal{F}_0$ of $\mathcal{F}$ for a fixed nonempty set X such that

- Each member of this finite sub-collection has the separation property for an intended purpose.
- This finite sub-collection $\mathcal{F}_0$ is a chain for the induced set inclusion.

Definition 1.42. Let $(X, \mathcal{T})$ be a topological space.

(i) X is called a T_1-*space* if for each pair of distinct elements $x, y \in X$, there are open subsets U and V of X such that $x \in U$, $y \notin U$, $y \in V$, and $x \notin V$.

(ii) X is called a T_2-*space* or a *Hausdorff space* if for each pair of distinct elements $x, y \in X$, there exist open subsets U and V of X such that $x \in U$, $y \in V$, and $U \cap V = \emptyset$.

(iii) X is called a *regular space* if for any closed subset K of X and any $x \in X \setminus K$, there exist open subsets U and V of X such that $K \subset U$, $x \in V$, and $U \cap V = \emptyset$.

(*iv*) X is called a T_3*-space* if X is a T_1-space and a regular space.

(*v*) X is called a *normal space* if for each pair of disjoint closed subsets H and K of X, there exist open subsets U and V of X such that $H \subset U$, $K \subset V$, and $U \cap V = \emptyset$.

(*vi*) X is called a T_4*-space* if X is a T_1-space and a normal space.

Let $(X, \mathcal{T})$ be a T_1-space and let x be an arbitrary element of X. According to Definition 1.42 (i), for each element y of $X \setminus \{x\}$, there exists an open subset V_y of X such that $y \in V_y$ but $x \notin V_y$. Then, it holds that $X \setminus \{x\} = \bigcup_{y \neq x} V_y \in \mathcal{T}$. Hence, $\{x\} = X \setminus \bigcup_{y \neq x} V_y$ is closed in X. That is, every set consisting of single element of the T_1-space X is closed in X. Conversely, if each set of single element of X is closed in X, then X is a T_1-space.

Let $X = \{a, b, c\}$ be a set of three elements. Then we may verify that $\mathcal{T} = \{\emptyset, \{a\}, \{b\}, \{a, b\}, X\}$ is a topology for X. We now see that $X \setminus \{c\} = \{a, b\} \in \mathcal{T}$. Thus, $\{c\}$ is closed in $(X, \mathcal{T})$. On the other hand, the sets $\{a\}$ and $\{b\}$ are not closed in $(X, \mathcal{T})$ because $X \setminus \{a\}$ and $X \setminus \{b\}$ do not belong to $\mathcal{T}$. This example shows that there is also a topological space that is not a T_1-space.

We assert that in each Hausdorff space, the limits of sequences are uniquely determined: let $\{x_i\}_{i \in \mathbb{N}}$ be a sequence in a Hausdorff space X that converges to x and y for some $x, y \in X$. Assume that $x \neq y$. According to Definition 1.42 (ii), there exist open subsets U and V of X such that $x \in U$, $y \in V$, and $U \cap V = \emptyset$. Since the sequence $\{x_i\}_{i \in \mathbb{N}}$ converges to x, there exists an $i_0 \in \mathbb{N}$ such that $x_i \in U$ for all $i \geq i_0$. On the other hand, it follows from the disjointness of U and V that $x_i \notin V$ for all $i \geq i_0$, which is contrary to the assumption that $\{x_i\}_{i \in \mathbb{N}}$ converges to y. Therefore, we conclude that $x = y$.

Assume that (X, d) is a metric space and $\mathcal{T}(d)$ is the topology for X generated by d. We now assert that the topological space $(X, \mathcal{T}(d))$ is a Hausdorff space: let x and y be distinct elements of X. Then we have $\delta := d(x, y) > 0$. If we define the open sets $U = B_{(1/2)\delta}(x)$ and $V = B_{(1/2)\delta}(y)$, then $x \in U$, $y \in V$, and $U \cap V = \emptyset$. Therefore, we conclude that $(X, \mathcal{T}(d))$ is a Hausdorff space.

Not every compact subset of a topological space needs to be closed. However, as shown in the following theorem, all compact subsets of any Hausdorff space are closed subsets.

Theorem 1.43. *Every compact subset of a Hausdorff space X is a closed subset of X.*

Proof. We assume that K is a nonempty compact subset of a Hausdorff space X, without loss of generality. Let x be an arbitrary element of $X \setminus K$. Since X is a

Hausdorff space, for every $y \in K$, there exist disjoint open subsets $U(y)$ and $V(y)$ of X such that $x \in U(y)$ and $y \in V(y)$. Then the collection $\{V(y) : y \in K\}$ is an open covering of K. Since K is compact, there exists a finite sub-covering $\{V(y_1), V(y_2), \ldots, V(y_n)\}$. If we set $U = \bigcap_{i=1}^{n} U(y_i)$ and $V = \bigcup_{i=1}^{n} V(y_i)$, then U and V are disjoint open subsets of X satisfying $x \in U$ and $K \subset V$. Since $x \in U \subset X \setminus K$, it holds that $X \setminus K$ is open in X, and hence, K is closed in X.
□

The idea that was useful in proving Theorem 1.43 will also be used to prove the following theorem.

Theorem 1.44. *Assume that H and K are disjoint compact subsets of a Hausdorff space X. Then there exist disjoint open subsets U and V of X such that $H \subset U$ and $K \subset V$.*

Proof. Let x be a fixed element of H and let y be an arbitrary element of K. Since X is a Hausdorff space, there are open subsets $U_x(y)$ and $V_x(y)$ of X such that $x \in U_x(y)$, $y \in V_x(y)$, and $U_x(y) \cap V_x(y) = \emptyset$.

The collection $\{V_x(y) : y \in K\}$ is an open covering of the compact set K. Hence, there exists a finite sub-collection $\{V_x(y_1), V_x(y_2), \ldots, V_x(y_m)\}$ that also covers K. We now define $U_x = \bigcap_{i=1}^{m} U_x(y_i)$ and $V_x = \bigcup_{i=1}^{m} V_x(y_i)$. Then, U_x and V_x are open subsets of X such that $x \in U_x$, $K \subset V_x$, and $U_x \cap V_x = \emptyset$.

On the other hand, the collection $\{U_x : x \in H\}$ is an open covering of the compact set H. Thus, there exists a finite sub-collection $\{U_{x_1}, U_{x_2}, \ldots, U_{x_n}\}$ that also covers H. We define $U = \bigcup_{i=1}^{n} U_{x_i}$ and $V = \bigcap_{i=1}^{n} V_{x_i}$. Then, U and V are open subsets of X such that $H \subset U$, $K \subset V$, and $U \cap V = \emptyset$. □

1.7 Open and Closed Subsets

In Definition 1.4, we defined the relative topology for a subset of a topological space. We will show in the next lemma that we can make a similar claim for closed subsets.

Lemma 1.45. *Assume that Y is a subset of a topological space X and W is a subset of Y.*

(*i*) *W is open in Y if and only if there exists an open subset U of X such that $W = U \cap Y$.*

(ii) *W is closed in Y if and only if there exists a closed subset K of X such that $W = K \cap Y$.*

Proof. (i) Since this statement is only the definition of the open set in the topological subspace, there is nothing to prove.

(ii) If W is closed in Y, then $Y \setminus W$ is open in Y. Thus, there exists an open subset U of X that satisfies $Y \setminus W = U \cap Y$. That is, $W = Y \setminus (Y \setminus W) = Y \setminus (U \cap Y) = Y \setminus U = Y \cap (X \setminus U)$, where $X \setminus U$ is a closed subset of X. On the other hand, if there exists a closed subset K of X that satisfies $W = K \cap Y$, then we have $Y \setminus W = Y \setminus (K \cap Y) = Y \setminus K = Y \cap (X \setminus K)$, where $X \setminus K$ is an open subset of X. Hence, $Y \setminus W$ is open in Y, i.e., W is closed in Y. □

We will find that Lemma 1.45 is particularly useful for proving Theorem 1.46.

Theorem 1.46. *Assume that Y is a subset of a topological space $(X, \mathcal{T})$ and W is a subset of Y.*

(i) *Assume that Y is open in X. Then W is open in Y if and only if W is open in X.*

(ii) *Assume that Y is closed in X. Then W is closed in Y if and only if W is closed in X.*

Proof. (i) Assume that $Y \in \mathcal{T}$. According to Lemma 1.45, for any $W \in \mathcal{T}_Y$, there exists a $U \in \mathcal{T}$ such that $W = U \cap Y$, i.e., W is open in X. On the other hand, assume that $W \in \mathcal{T}$. Then, $W = W \cap Y \in \mathcal{T}_Y$, i.e., W is open in Y.

(ii) Assume that Y is closed in X and W is closed in Y. Due to Lemma 1.45, there exists a closed subset K of X that satisfies $W = K \cap Y$ which is clearly closed in X. On the other hand, assume that W is closed in X. Then, due to Lemma 1.45, $W = W \cap Y$ is closed in Y. □

The following three theorems are directly quoted from [12]. The next theorem deals with a necessary and sufficient condition that an open subset of a closed subset of a topological space X becomes an open subset of X.

Theorem 1.47. *Assume that Y is a closed subset of a topological space X and W is an open subset of Y. Then W is open in X if and only if $W \subset Y^\circ$.*

Proof. Assume that W is open in X. Since Y° is the union of all open subsets of X which are contained in Y, it follows that $W \subset Y^\circ$. Since W is open in Y, Lemma 1.45 implies that there exists an open subset U of X satisfying $W = U \cap Y$. If $W \subset Y^\circ$, then

$$W = W \cap Y^\circ = U \cap Y \cap Y^\circ = U \cap Y^\circ.$$

Since U and Y° is open in X, we conclude that W is open in X. □

The following theorem provides the necessary and sufficient condition that a closed subset of the open subspace of a topological space X becomes a closed subset of X.

Theorem 1.48. *Assume that Y is an open subset of a topological space X and W is a closed subset of Y. Then W is closed in X if and only if $\overline{W} \subset Y$.*

Proof. If W is closed in X, then it is obvious that $W = \overline{W} \subset Y$. On the other hand, since W is closed in Y, by Lemma 1.45, there exists a closed subset K of X satisfying $W = K \cap Y$. If $\overline{W} \subset Y$, then

$$W = W \cap \overline{W} = K \cap Y \cap \overline{W} = K \cap \overline{W}.$$

Since K and $\overline{W}$ are closed in X, we conclude that W is closed in X. □

Obviously, $U \cup V$ is an open set whenever both U and V are open sets. Conversely, if $U \cup V$ is open, under what conditions can we expect that both U and V are open?

Theorem 1.49. *Assume that U and V are arbitrary subsets of a topological space X which are mutually separated in X, i.e., $\overline{U} \cap V = U \cap \overline{V} = \emptyset$. If $U \cup V$ is open, then both U and V are open.*

Proof. Since U and V are mutually separated, it holds that $U \subset X \backslash \overline{V}$. Since $U \cup V$ and $X \backslash \overline{V}$ are open, their intersection

$$(U \cup V) \cap (X \backslash \overline{V}) = (U \cap X \backslash \overline{V}) \cup (V \cap X \backslash \overline{V}) = U \cup \emptyset = U$$

is open. In the same way, we can prove that V is open. □

Chapter 2

Hilbert Spaces

In this chapter, we briefly introduce basic concepts and theorems in vector space, normed space, Banach space, and Hilbert space that are essential to prove Ulam's conjecture, the main subject of this book. For this purpose, we mainly refer to the book [3] by L. Debnath and P. Mikusiński, among many other literatures. In the following chapters, we will see that some properties of the inner product are essential.

2.1 Vector Spaces

Let $\mathbb{R}$ be the field of all real numbers and let $\mathbb{C}$ be the field of all complex numbers.

Definition 2.1. Let $\mathbb{K}$ denote either $\mathbb{R}$ or $\mathbb{C}$. A nonempty set X is called a *vector space* over $\mathbb{K}$ if there are two operations, called *addition* and *scalar multiplication*, such that the following conditions are satisfied:

(i) $x + y = y + x$ for all $x, y \in X$.

(ii) $(x + y) + z = x + (y + z)$ for all $x, y, z \in X$.

(iii) For any $x, y \in X$, there exists a $z \in X$ such that $x + z = y$.

(iv) $\alpha(\beta x) = (\alpha\beta)x$ for all $\alpha, \beta \in \mathbb{K}$ and any $x \in X$.

(v) $(\alpha + \beta)x = \alpha x + \beta x$ for all $\alpha, \beta \in \mathbb{K}$ and any $x \in X$.

(vi) $\alpha(x + y) = \alpha x + \alpha y$ for all $\alpha \in \mathbb{K}$ and all $x, y \in X$.

(vii) $1x = x$ for all $x \in X$.

S.-M. Jung, *Ulam's Conjecture on Invariance of Measure in the Hilbert Cube*, Frontiers in Mathematics, https://doi.org/10.1007/978-3-031-30886-4_2

Elements of $\mathbb{K}$ are called *scalars* and elements of X are called *vectors*. If $\mathbb{K} = \mathbb{R}$, then X is called a *real vector space*, and if $\mathbb{K} = \mathbb{C}$, then X is called a *complex vector space*.

A subset Y of a vector space X over $\mathbb{K}$ is called a *vector subspace* or a *subspace* if $\alpha x + \beta y \in Y$ for all $\alpha, \beta \in \mathbb{K}$ and $x, y \in Y$. We note that any subspace of a vector space is a vector space itself and each vector space is a subspace of itself. A subset Y of a vector space X is called a *proper subspace* of X if Y is a subspace of X and $Y \neq X$.

Throughout this book, we use the notation $\{x_i\}_{i \in \mathbb{N}}$ or $\{x_1, x_2, \ldots\}$ to denote the sequence whose ith term is x_i.

Let X be the set of all sequences $\{x_i\}_{i \in \mathbb{N}}$ of real numbers. We define the addition and scalar multiplication by

$$\{x_i\}_{i \in \mathbb{N}} + \{y_i\}_{i \in \mathbb{N}} = \{x_i + y_i\}_{i \in \mathbb{N}},$$
$$\alpha \{x_i\}_{i \in \mathbb{N}} = \{\alpha x_i\}_{i \in \mathbb{N}}$$

for all $\alpha \in \mathbb{R}$ and $\{x_i\}_{i \in \mathbb{N}}, \{y_i\}_{i \in \mathbb{N}} \in X$. Then X is a real vector space. The space of all bounded sequences of real numbers is a proper subspace of X. The space of all convergent sequences of real numbers is a proper subspace of the space of all bounded sequences.

2.2 Basis of Vector Space

Let $x_1, x_2, \ldots, x_n$ be elements of a vector space X. An element x of X is called a *linear combination* of vectors $x_1, x_2, \ldots, x_n$ if there exist scalars $\alpha_1, \alpha_2, \ldots, \alpha_n$ such that

$$x = \alpha_1 x_1 + \alpha_2 x_2 + \cdots + \alpha_n x_n.$$

For example, each element of the n-dimensional Euclidean space $\mathbb{R}^n$ is a linear combination of vectors $e_1, e_2, \ldots, e_n$, where we set $e_i = (0, \ldots, 0, 1, 0, \ldots, 0)$ with 1 in the ith position.

Definition 2.2. Let X be a vector space.

(i) A finite collection $\{x_1, x_2, \ldots, x_n\}$ of elements of X is called *linearly independent* if $\alpha_1 = \alpha_2 = \cdots = \alpha_n = 0$ is the unique solution of the linear equation $\alpha_1 x_1 + \alpha_2 x_2 + \cdots + \alpha_n x_n = 0$.

(ii) An infinite collection $\mathcal{A}$ of elements of X is called *linearly independent* if every finite sub-collection of $\mathcal{A}$ is linearly independent.

(*iii*) A collection of elements of X is called *linearly dependent* if it is not linearly independent.

We note that a collection $\mathcal{A}$ of elements of X is linearly independent if no member x of the collection $\mathcal{A}$ is a linear combination of a finite number of members of $\mathcal{A}$ different from x.

Definition 2.3. Let A be a subset of a vector space X over $\mathbb{K}$. We denote by span$A$ the set of all finite linear combinations of vectors from A, i.e.,

$$\text{span}A = \{\alpha_1 x_1 + \alpha_2 x_2 + \cdots + \alpha_n x_n : n \in \mathbb{N},\ \alpha_i \in \mathbb{K},\ x_i \in A \\ \text{for all } i \in \{1, 2, \ldots, n\}\}.$$

Then span$A$ is a vector subspace of X and it is called the *space spanned by A*.

It is easy to see that span$A$ is the smallest vector subspace of X containing A.

Definition 2.4. Let B be a subset of a vector space X. Then B is called a *basis* of X if B is linearly independent and span$B = X$.

In general, a vector space X has multiple bases, but the number of vectors in each basis is the same. In other words, if a basis of X has exactly n vectors, then every other basis also has exactly n vectors. In this case n is called the *dimension* of X and we write dim$X = n$.

2.3 Normed Spaces

The concept of the norm in the vector spaces is an abstract generalization of length in the Euclidean space. That is, the norm is axiomatically defined as a real-valued function that satisfies certain conditions.

Definition 2.5. Let X be a vector space over $\mathbb{K}$. A function $\|\cdot\| : X \to \mathbb{R}$ is called a *norm* if it satisfies the following conditions:

(*i*) $\|x\| = 0$ if and only if $x = 0$.

(*ii*) $\|\alpha x\| = |\alpha|\|x\|$ for all $x \in X$ and $\alpha \in \mathbb{K}$.

(*iii*) $\|x + y\| \leq \|x\| + \|y\|$ for all $x, y \in X$. In particular, this inequality is called the *triangle inequality*.

Since $\|x\| = \frac{1}{2}(\|x\| + \|-x\|) \geq \frac{1}{2}\|x + (-x)\| = \frac{1}{2}\|0\| = 0$, it holds that $\|x\| \geq 0$ for any element x of a vector space X with the norm $\|\cdot\|$.

The function $\|\cdot\| : \mathbb{R}^n \to \mathbb{R}$ defined by $\|x\| = \sqrt{x_1^2 + x_2^2 + \cdots + x_n^2}$, for all $x = (x_1, x_2, \ldots, x_n) \in \mathbb{R}^n$, is a norm on $\mathbb{R}^n$. This norm is called the *Euclidean norm*.

Definition 2.6. A vector space with a norm is called a *normed space*.

Several different norms can be defined on a vector space. For example, if we define $\|x\|_1 = |x_1| + |x_2| + \cdots + |x_n|$ for all $x = (x_1, x_2, \ldots, x_n) \in \mathbb{R}^n$, then this function $\|\cdot\|_1 : \mathbb{R}^n \to \mathbb{R}$ is a norm on the vector space $\mathbb{R}^n$. Similarly, we can see that the function $\|x\|_\infty = \max\{|x_1|, |x_2|, \ldots, |x_n|\}$ is also a norm on the vector space $\mathbb{R}^n$. This norm $\|\cdot\|_\infty$ is called the *sup-norm.*

Therefore, to define a normed space, we have to specify both the vector space and the norm. We denote a normed space as $(X, \|\cdot\|)$, where X is a vector space and $\|\cdot\|$ is a norm defined on X. However, if we are certain of the norm given in the norm space $(X, \|\cdot\|)$, we can simply write that norm space as X.

Definition 2.7. Let $(X, \|\cdot\|)$ be a normed space. A sequence $\{x_i\}_{i\in\mathbb{N}}$ of elements of X is said to *converge* to an $x \in X$ if for each $\varepsilon > 0$ there exists a positive integer N such that $\|x_i - x\| < \varepsilon$ for all integers $i \geq N$. In this case, we write $\lim_{i\to\infty} x_i = x$ or simply $x_i \to x$.

We note that $x_i \to x$ in X implies $\|x_i - x\| \to 0$ in $\mathbb{R}$. As we see in the following remark, the convergence of sequences in the normed space has the basic properties of the convergence of sequences in $\mathbb{R}$.

Remark 2.8. Let $(X, \|\cdot\|)$ be a normed space over $\mathbb{K}$ and let $\{x_i\}_{i\in\mathbb{N}}$ and $\{y_i\}_{i\in\mathbb{N}}$ be sequences of elements of X.

(i) If $\{x_i\}_{i\in\mathbb{N}}$ converges in X, then it has a unique limit.

(ii) If $x_i \to x$ and $\lambda_i \to \lambda$, where $x \in X$ and $\lambda_i, \lambda \in \mathbb{K}$, then $\lambda_i x_i \to \lambda x$.

(iii) If $x_i \to x$ and $y_i \to y$ for some $x, y \in X$, then $x_i + y_i \to x + y$.

As we see in Definition 2.7, every norm on a vector space X induces a convergence in X. But the reverse does not hold. For example, let $\mathcal{C}([0, 1])$ be the set of all continuous complex-valued functions defined on $[0, 1]$. Then it is known that $\mathcal{C}([0, 1])$ is a complex vector space and there is no norm inducing the pointwise convergence for this space.

Definition 2.9. Let $(X, \|\cdot\|_1)$ and $(X, \|\cdot\|_2)$ be normed spaces over $\mathbb{K}$. The norms $\|\cdot\|_1$ and $\|\cdot\|_2$ are called *equivalent* if they induce the same convergence. More precisely, the norms $\|\cdot\|_1$ and $\|\cdot\|_2$ are equivalent if

$$\|x_i - x\|_1 \to 0 \quad \text{if and only if} \quad \|x_i - x\|_2 \to 0$$

for all sequences $\{x_i\}_{i\in\mathbb{N}}$ in X and $x \in X$.

It is well known that any two norms on a finite-dimensional vector space are equivalent.

The following theorem gives another useful criterion for the equivalence of norms. The condition in the theorem is often used as a definition of the equivalence of norms.

Theorem 2.10. *Let $\|\cdot\|_1$ and $\|\cdot\|_2$ be norms on a vector space X. Then the norms $\|\cdot\|_1$ and $\|\cdot\|_2$ are equivalent if and only if there exist positive real numbers α and β such that*

$$\alpha\|x\|_1 \leq \|x\|_2 \leq \beta\|x\|_1 \tag{2.1}$$

for all $x \in X$.

Proof. Assume that the norms $\|\cdot\|_1$ and $\|\cdot\|_2$ are equivalent. Then, it follows from Definition 2.9 that

$$\|x_i - x\|_1 \to 0 \quad \text{if and only if} \quad \|x_i - x\|_2 \to 0 \tag{2.2}$$

for all sequences $\{x_i\}_{i\in\mathbb{N}}$ in X and $x \in X$.

On the contrary, assume that for any $\alpha > 0$ there is an $x \in X$ such that $\alpha\|x\|_1 > \|x\|_2$. Then there exists an $x_i \in X$ that satisfies

$$\frac{1}{i}\|x_i\|_1 > \|x_i\|_2$$

for every $i \in \mathbb{N}$. We set

$$y_i = \frac{1}{\sqrt{i}}\frac{1}{\|x_i\|_2}x_i$$

for each $i \in \mathbb{N}$. It then holds that $\|y_i\|_2 = \frac{1}{\sqrt{i}} \to 0$ as $i \to \infty$. On the other hand, it holds that $\|y_i\|_1 \geq \sqrt{i} \to \infty$ as $i \to \infty$, which is contrary to (2.2). Therefore, we conclude that there exists an $\alpha > 0$ such that $\alpha\|x\|_1 \leq \|x\|_2$ for all $x \in X$. In a similar way, we can prove the existence of the positive number β.

By Definition 2.9, it is obvious that the condition (2.1) implies the equivalence of the norms $\|\cdot\|_1$ and $\|\cdot\|_2$. □

Let $(X, \|\cdot\|)$ be a normed space. If we define a metric by $d(x, y) = \|x - y\|$, then (X, d) becomes a metric space. The convergence induced by the norm is the same as the convergence induced by this metric. In view of Definition 1.5, the metric on X generates a topology for X. Now that we can see the normed space as a topological space, we will use the norm to define basic topological concepts.

Since every normed space is a metric space, we would not get confused even if we use in the following definition the same symbols that we used for the metric space.

Definition 2.11. Let x be an element of a normed space $(X, \|\cdot\|)$ and let r be a positive real number.

(i) We denote by $B_r(x)$ the *open ball*, with center x and radius r, defined by $B_r(x) = \{y \in X : \|y - x\| < r\}$.

(ii) We denote by $\overline{B}_r(x)$ the *closed ball*, with center x and radius r, defined by $\overline{B}_r(x) = \{y \in X : \|y - x\| \leq r\}$.

(iii) Similarly, we use the symbol $S_r(x)$ to denote the *sphere*, with center x and radius r, defined by $S_r(x) = \{y \in X : \|y - x\| = r\}$.

Since, as already mentioned, every normed space is a metric space and thus every normed space is a topological space, Lemma 1.3 also holds literally for every normed space.

Lemma 2.12. *Assume that x is an element of a normed space $(X, \|\cdot\|)$ and r is a positive real number.*

(i) *The open ball $B_r(x)$ is open in X.*

(ii) *The closed ball $\overline{B}_r(x)$ is closed in X.*

(iii) *The sphere $S_r(x)$ is closed in X.*

We recall that every normed space is a metric space.

Theorem 2.13. *Each compact subset of a normed space is closed and bounded.*

Proof. Let K be a compact subset of a normed space $(X, \|\cdot\|)$. Assume that $\{x_i\}_{i \in \mathbb{N}}$ is a sequence of elements of K that converges to an element x of X. Since the compactness is equivalent to the sequential compactness for any metric space (see Sect. 1.2), the sequence $\{x_i\}_{i \in \mathbb{N}}$ contains a subsequence $\{x_{p_i}\}_{i \in \mathbb{N}}$ that converges to some element y of K. Moreover, the subsequence $\{x_{p_i}\}_{i \in \mathbb{N}}$ converges to x. That is, $x = y$ and $x \in K$, which implies that K is closed.

Now we assume that K is not bounded. Then there exists a sequence $\{x_i\}_{i \in \mathbb{N}}$ of elements of K such that $\|x_i\| \geq i$ for each $i \in \mathbb{N}$. It is obvious that the sequence $\{x_i\}_{i \in \mathbb{N}}$ does not contain a convergent subsequence. Hence, K is not sequentially compact, i.e., K is not compact, which leads to a contradiction. Therefore, K has to be bounded. □

We note that a subset of a finite-dimensional normed space is compact if and only if it is closed and bounded. However, this is not generally the case.

Example 2.14. We consider the normed space $\mathcal{C}([0,1])$. The unit closed ball $\overline{B}_1(0)$ is a closed and bounded subset of $\mathcal{C}([0,1])$, but it is not compact. To show this, we consider the sequence of functions defined by $x_i(t) = t^i$ for all $i \in \mathbb{N}$. Then $x_i \in \overline{B}_1(0)$ for all $i \in \mathbb{N}$. Since the convergence in $\mathcal{C}([0,1])$ is the uniform convergence, the sequence $\{x_i\}_{i\in\mathbb{N}}$ does not have a convergent subsequence.

2.4 Banach Spaces

We note that every Cauchy sequence of elements from $\mathbb{K}^n$ converges and every absolutely convergent series of elements from $\mathbb{K}^n$ converges. However, not all normed spaces have the above properties.

We recall that a sequence $\{x_i\}_{i\in\mathbb{N}}$ of vectors in a normed space $(X, \|\cdot\|)$ is called a *Cauchy sequence* if for each $\varepsilon > 0$ there exists a positive real number M_ε such that $\|x_i - x_j\| < \varepsilon$ for all $i, j > M_\varepsilon$.

If a sequence $\{x_i\}_{i\in\mathbb{N}}$ of vectors in a normed space $(X, \|\cdot\|)$ converges to a vector $x \in X$, i.e., $\|x_i - x\| \to 0$ as $i \to \infty$, then

$$\|x_i - x_j\| \le \|x_i - x\| + \|x_j - x\| \to 0$$

as $i, j \to \infty$. This explanation implies that every convergent sequence of vectors in a normed space is a Cauchy sequence. However, the converse does not hold in general. For example, we denote by $\mathcal{P}([0,1])$ the normed space of polynomials on $[0,1]$, where the norm is given by $\|P\| = \max\limits_{0\le x\le 1} |P(x)|$. We define

$$P_i(x) = 1 + x + \frac{1}{2!}x^2 + \cdots + \frac{1}{i!}x^i$$

for all $i \in \mathbb{N}$. Then $\{P_i\}_{i\in\mathbb{N}}$ is a Cauchy sequence. However, it does not converge in $\mathcal{P}([0,1])$ since its limit is not a polynomial.

Lemma 2.15. *If $\{x_i\}_{i\in\mathbb{N}}$ is a Cauchy sequence in a normed space $(X, \|\cdot\|)$, then the sequence converges.*

Proof. Since $|\|x_i\| - \|x_j\|| \le \|x_i - x_j\|$ for all $i, j \in \mathbb{N}$, the sequence $\{\|x_i\|\}_{i\in\mathbb{N}}$ is a Cauchy sequence in $\mathbb{R}$. Hence, we conclude that $\{\|x_i\|\}_{i\in\mathbb{N}}$ converges. □

Lemma 2.15 implies that each Cauchy sequence in a normed space is bounded.

Definition 2.16. Let $(X, \|\cdot\|)$ be a normed space.

(i) X is called *complete* if each Cauchy sequence in X converges to an element of X.

(*ii*) If X is a complete normed space, then it is called a *Banach space*.

The n-dimensional Euclidean space $\mathbb{R}^n$ is a typical Banach space.

Definition 2.17. Let $(X, \|\cdot\|)$ be a normed space.

(*i*) A series $\sum_{i=1}^{\infty} x_i$ *converges* in X if there exists an element x of X such that

$$\left\| \sum_{i=1}^{n} x_i - x \right\| \to 0 \quad \text{as} \quad n \to \infty.$$

In this case, we write $\sum_{i=1}^{\infty} x_i = x$.

(*ii*) When $\sum_{i=1}^{\infty} \|x_i\| < \infty$, the series is called *absolutely convergent*.

In general, not all series converge even if they are absolutely convergent. The following theorem shows that completeness and absolute convergence of series are equivalent.

Theorem 2.18. *A normed space $(X, \|\cdot\|)$ is complete if and only if every absolutely convergent series in X converges in X.*

Proof. Assume that $(X, \|\cdot\|)$ is a complete normed space and $x_i \in X$ for all $i \in \mathbb{N}$ such that $\sum_{i=1}^{\infty} \|x_i\| < \infty$, i.e., the series $\sum_{i=1}^{\infty} x_i$ converges absolutely. Let us define

$$s_n = x_1 + x_2 + \cdots + x_n$$

for all $n \in \mathbb{N}$.

We prove that $\{s_n\}_{n\in\mathbb{N}}$ is a Cauchy sequence. Assume that ε is an arbitrary positive real number and N is a positive integer that satisfies

$$\sum_{n=N+1}^{\infty} \|x_n\| < \varepsilon.$$

Then, by triangle inequality, we get

$$\|s_m - s_n\| = \|x_{m+1} + x_{m+2} + \cdots + x_n\| \leq \sum_{k=m+1}^{\infty} \|x_k\| < \varepsilon$$

for all $m, n \in \mathbb{N}$ with $n > m > N$, which implies that $\{s_n\}_{n\in\mathbb{N}}$ is a Cauchy sequence in X.

Since X is complete, the Cauchy sequence $\{s_n\}_{n\in\mathbb{N}}$ converges in X, which implies that the series $\sum\limits_{i=1}^{\infty} x_i$ converges in X.

Now we assume that every absolutely convergent series converges and $\{x_i\}_{i\in\mathbb{N}}$ is an arbitrary Cauchy sequence in X. Then, for any positive integer k, there exists a positive integer p_k such that $\|x_i - x_j\| < \frac{1}{2^k}$ for all integers $i, j \geq p_k$. We may assume that the sequence $\{p_k\}_{k\in\mathbb{N}}$ is strictly increasing.

Since the series $\sum\limits_{k=1}^{\infty}(x_{p_{k+1}} - x_{p_k})$ is absolutely convergent, it is convergent by our assumption. Since

$$x_{p_k} = x_{p_1} + (x_{p_2} - x_{p_1}) + \cdots + (x_{p_k} - x_{p_{k-1}}) = x_{p_1} + \sum_{i=1}^{k}(x_{p_{i+1}} - x_{p_i}),$$

the sequence $\{x_{p_k}\}_{k\in\mathbb{N}}$ converges to an element x of X. Hence, it holds that

$$\|x_i - x\| \leq \|x_i - x_{p_i}\| + \|x_{p_i} - x\| \to 0$$

as $i \to \infty$, i.e., the Cauchy sequence $\{x_i\}_{i\in\mathbb{N}}$ converges in X, which implies that X is complete. □

Theorem 2.19. *A closed subspace of a Banach space is a Banach space.*

Proof. Let K be a closed subspace of a Banach space X. Then K is a normed space which is closed in X. Due to Theorem 1.41, K is complete as a closed subset of a complete space X. Therefore, K is a Banach space. □

2.5 Inner Product Spaces

An inner product space is a vector space over $\mathbb{K}$ with an operation called an inner product. The inner product of two vectors x and y in the space is a scalar, denoted with angle brackets such as $\langle x, y\rangle$. Inner products allow formal definitions of intuitive geometric notions, such as lengths, angles, and orthogonality of vectors. For this reason, the infinite-dimensional inner product spaces are widely used in functional analysis.

We use the symbol $\overline{z}$ to denote the complex conjugation of any complex number z.

Definition 2.20. Let X be a vector space over $\mathbb{K}$. A function $\langle\cdot,\cdot\rangle : X \times X \to \mathbb{K}$ is called an *inner product* on X if the function satisfies the conditions

(*i*) $\langle x, y\rangle = \overline{\langle y, x\rangle}$

(*ii*) $\langle \alpha x + \beta y, z \rangle = \alpha \langle x, z \rangle + \beta \langle y, z \rangle$

(*iii*) $\langle x, x \rangle \geq 0$

(*iv*) $\langle x, x \rangle = 0$ if and only if $x = 0$

for all $x, y, z \in X$ and $\alpha, \beta \in \mathbb{K}$. A vector space with an inner product is called an *inner product space* or a *pre-Hilbert space*.

It follows from Definition 2.20 (i) that $\langle x, x \rangle = \overline{\langle x, x \rangle}$, which implies that $\langle x, x \rangle$ is a real number for any $x \in X$.

Due to Definition 2.20 (i) and (ii), we have

$$\langle x, \alpha y + \beta z \rangle = \overline{\langle \alpha y + \beta z, x \rangle} = \overline{\alpha} \langle x, y \rangle + \overline{\beta} \langle x, z \rangle$$

for all $x, y, z \in X$ and $\alpha, \beta \in \mathbb{K}$.

Example 2.21. Let $\mathbb{K}^n$ be the vector space of all ordered n-tuples $(x_1, x_2, \ldots, x_n)$, where $x_i \in \mathbb{K}$ for each $i \in \{1, 2, \ldots, n\}$. We define a function $\langle \cdot, \cdot \rangle : \mathbb{K}^n \times \mathbb{K}^n \to \mathbb{K}$ by

$$\langle x, y \rangle = \sum_{i=1}^{n} x_i \overline{y}_i$$

for all $x = (x_1, x_2, \ldots, x_n)$ and $y = (y_1, y_2, \ldots, y_n)$. Then $\langle \cdot, \cdot \rangle$ is an inner product on the vector space $\mathbb{K}^n$ and $(\mathbb{K}^n, \langle \cdot, \cdot \rangle)$ is an inner product space.

Every inner product $\langle \cdot, \cdot \rangle$ on a vector space X naturally induces an associated norm through $\|x\| = \sqrt{\langle x, x \rangle}$ for all $x \in X$. The reader is encouraged to prove, as an exercise, that the function defined above is indeed a norm. Therefore, every inner product space is a normed space.

In general, the norm on an inner product space means the function defined by $\|x\| = \sqrt{\langle x, x \rangle}$.

Theorem 2.22 (Cauchy–Schwarz Inequality). *Let $(X, \langle \cdot, \cdot \rangle)$ be an inner product space. Then*

$$|\langle x, y \rangle| \leq \|x\| \|y\|$$

for all $x, y \in X$.

Proof. For any $x, y \in X$, we set $A = \|x\|^2 = \langle x, x \rangle$, $B = |\langle x, y \rangle|$, and $C = \|y\|^2 = \langle y, y \rangle$. We note that B is a nonnegative real number. Then we can choose a scalar $\alpha \in \mathbb{K}$ such that $|\alpha| = 1$ and $B = \alpha\langle y, x\rangle$. For all $r \in \mathbb{R}$, it follows from Definition 2.20 that

$$\begin{aligned} 0 &\leq \langle x - r\alpha y, x - r\alpha y\rangle \\ &= \langle x, x\rangle - r\alpha\langle y, x\rangle - r\overline{\alpha}\langle x, y\rangle + r^2\langle y, y\rangle \\ &= \langle x, x\rangle - r\alpha\langle y, x\rangle - r\overline{\alpha\langle y, x\rangle} + r^2\langle y, y\rangle \\ &= A - Br - \overline{B}r + r^2\langle y, y\rangle \\ &= A - 2Br + Cr^2. \end{aligned}$$

If $C = 0$, it has to be $B = 0$. (Otherwise, the above inequality does not hold for large $r > 0$.) If $C > 0$, we take $r = \frac{B}{C}$ in the above inequality and we obtain $B^2 \leq AC$, which completes the proof. □

It is a natural thing to ask whether every normed space is an inner product space. Unfortunately, the answer is negative. In the following theorem, we propose a condition for a norm on an inner product space, which is a necessary and sufficient condition for the normed space to be an inner product space.

We recall that the norm on an inner product space is defined by $\|x\| = \sqrt{\langle x, x\rangle}$.

Theorem 2.23 (Parallelogram Law). *Let $(X, \langle\cdot,\cdot\rangle)$ be an inner product space. Then*

$$\|x + y\|^2 + \|x - y\|^2 = 2\|x\|^2 + 2\|y\|^2 \tag{2.3}$$

for all $x, y \in X$.

Proof. It holds that

$$\begin{aligned} \|x + y\|^2 &= \langle x + y, x + y\rangle \\ &= \langle x, x\rangle + \langle x, y\rangle + \langle y, x\rangle + \langle y, y\rangle \\ &= \|x\|^2 + \langle x, y\rangle + \langle y, x\rangle + \|y\|^2 \end{aligned}$$

for all $x, y \in X$. If we replace y by $-y$ in the above equation, then we get

$$\|x - y\|^2 = \|x\|^2 - \langle x, y\rangle - \langle y, x\rangle + \|y\|^2$$

for all $x, y \in X$.

We obtain the parallelogram law by adding the last two equations. □

Let $(X, \|\cdot\|)$ be a normed space over $\mathbb{K}$. For any norm that satisfies the parallelogram law (2.3), the inner product that generates the norm is unique as a consequence of polarization identity. In the case of $\mathbb{K} = \mathbb{R}$, the polarization identity is given by

$$\langle x, y\rangle = \frac{1}{4}\left(\|x+y\|^2 - \|x-y\|^2\right) \tag{2.4}$$

for all $x, y \in X$. For the case of $\mathbb{K} = \mathbb{C}$, the polarization identity is given by

$$\langle x, y\rangle = \frac{1}{4}\left(\|x+y\|^2 - \|x-y\|^2\right) + \frac{i}{4}\left(\|ix-y\|^2 - \|ix+y\|^2\right) \tag{2.5}$$

for all $x, y \in X$.

So if the parallelogram law is satisfied in a normed space $(X, \|\cdot\|)$, then the normed space is an inner product space that is correspondingly equipped with the inner product (2.4) or (2.5).

Another necessary and sufficient condition for the existence of an inner product that induces the given norm is that the norm satisfies the Ptolemy inequality

$$\|x-y\|\|z\| + \|y-z\|\|x\| \geq \|z-x\|\|y\|.$$

More precisely, the Ptolemy inequality holds in every inner product space. Conversely, if the inequality holds in a real normed space, then the space has to be a real inner product space.

We will now prove that the inner product and norm are continuous functions using Cauchy–Schwarz inequality and the triangle inequality.

Theorem 2.24. *Let $(X, \langle\cdot,\cdot\rangle)$ be an inner product space over $\mathbb{K}$. For any fixed $x, y \in X$, the functions $\langle x, \cdot\rangle : X \to \mathbb{K}$, $\langle \cdot, y\rangle : X \to \mathbb{K}$, and $\|\cdot\| : X \to \mathbb{R}$ are (uniformly) continuous on X.*

Proof. By the Cauchy–Schwarz inequality, we have

$$|\langle x, y_1\rangle - \langle x, y_2\rangle| = |\langle x, y_1 - y_2\rangle| \leq \|x\|\|y_1 - y_2\|$$

for all $y_1, y_2 \in X$ and for a fixed $x \in X$. That is, $\langle x, \cdot\rangle$ is a uniformly continuous function. Similarly, we can prove that if y is a fixed element of X, then $\langle \cdot, y\rangle$ is a uniformly continuous function.

Using the triangle inequality, we obtain

$$\|x_1\| - \|x_2\| \leq \|x_1 - x_2\|,$$

and if we interchange x_1 and x_2 in the above inequality, then we see that

$$|\|x_1\| - \|x_2\|| \leq \|x_1 - x_2\|$$

for all $x_1, x_2 \in X$. Therefore, $\|\cdot\|$ is also a uniformly continuous function. □

We recall that a subset V of a vector space X is called a subspace of X if V is itself a vector space with the addition and scalar multiplication which are defined on X. In sentences relating to vector spaces, the term "subspace" always refers to the subspace mentioned above. Sometimes we write "linear subspace" instead of subspace.

Theorem 2.25. *If V is a subspace of an inner product space X, so is $\overline{V}$.*

Proof. Let x and y be arbitrary elements of $\overline{V}$ and let α be an arbitrary scalar. Then there exist sequences $\{x_i\}_{i\in\mathbb{N}}$ and $\{y_i\}_{i\in\mathbb{N}}$ in V which converge to x and y, respectively. Since V is a vector space, $\{x_i + y_i\}_{i\in\mathbb{N}}$ and $\{\alpha x_i\}_{i\in\mathbb{N}}$ are sequences in V which converge to $x + y$ and αx, respectively. Therefore, $x + y \in \overline{V}$ and $\alpha x \in \overline{V}$, which implies that $\overline{V}$ is a subspace of X. □

One of the most important consequences of the inner product is the ability to define the orthogonality of vectors. This makes the theory of Hilbert spaces very different from the general theory of Banach spaces.

Definition 2.26. Let X be an inner product space. Two vectors $x, y \in X$ are called *orthogonal* if $\langle x, y\rangle = 0$. In the case, we write $x \perp y$.

We note that if $x \perp y$, then $\langle x, y\rangle = 0$, and hence, $\langle y, x\rangle = \overline{\langle x, y\rangle} = \overline{0} = 0$, i.e., $y \perp x$.

Another example of the geometric character of the norm defined by an inner product is the Pythagorean theorem. The Pythagorean theorem is a fundamental relationship in Euclidean geometry between the three sides of a right triangle. It states that the area of the square whose side is the hypotenuse (the side opposite the right angle) is equal to the sum of the areas of the squares on the other two sides.

By the following theorem, the Pythagorean theorem holds in every inner product space.

Theorem 2.27 (Pythagorean Theorem). *Let $(X, \langle\cdot,\cdot\rangle)$ be an inner product space. If the norm $\|\cdot\|$ on X is induced by the inner product $\langle\cdot,\cdot\rangle$, then the Pythagorean formula holds, i.e.,*

$$\|x + y\|^2 = \|x\|^2 + \|y\|^2$$

for any pair of orthogonal vectors in X.

Proof. If x and y are orthogonal vectors, then $\langle x, y\rangle = \langle y, x\rangle = 0$. Hence, we have

$$\|x + y\|^2 = \|x\|^2 + \langle x, y\rangle + \langle y, x\rangle + \|y\|^2 = \|x\|^2 + \|y\|^2$$

for all $x, y \in X$. □

2.6 Hilbert Spaces

A Hilbert space is a vector space endowed with an inner product that induces a distance function for which it is a complete metric space. Because Hilbert spaces allow the methods of linear algebra and calculus to be generalized from (finite-dimensional) Euclidean spaces to possibly infinite-dimensional spaces, the Hilbert spaces are naturally and widely used in mathematics and physics.

Definition 2.28. A *Hilbert space* is a complete inner product space.

In the definition above, the completeness of an inner product space $(X, \langle \cdot, \cdot \rangle)$ means the completeness of the normed space $(X, \| \cdot \|)$, where the norm is defined by $\|x\| = \sqrt{\langle x, x \rangle}$ for all $x \in X$ (see Sect. 2.4).

There are many examples of Hilbert spaces. $\mathbb{R}^n$ and $\mathbb{C}^n$ are Hilbert spaces if they are equipped with the inner products $\langle x, y \rangle = \sum_{i=1}^{n} x_i y_i$ and $\langle x, y \rangle = \sum_{i=1}^{n} x_i \overline{y}_i$, respectively. Another example of Hilbert spaces is l^2, where l^2 is the space of all sequences $\{x_i\}_{i \in \mathbb{N}}$ of complex numbers such that $\sum_{i=1}^{\infty} |x_i|^2 < \infty$ with the inner product defined by

$$\langle x, y \rangle = \sum_{i=1}^{\infty} x_i \overline{y}_i.$$

We recall that two vectors in an inner product space are called orthonormal if they are orthogonal unit vectors. We say that a set of vectors forms an orthonormal set if all vectors in the set are mutually orthogonal and all have unit length.

Definition 2.29. Let $(X, \langle \cdot, \cdot \rangle)$ be an inner product space.

(i) A collection S of nonzero vectors in X is called an *orthogonal system* if any two different vectors in S are orthogonal to each other.

(ii) If every vector in an orthogonal system S is a unit vector, i.e., $\|x\| = 1$ for every $x \in S$, then S is called an *orthonormal system*.

We note that in Definition 2.29 (ii), the norm is induced by the inner product, i.e., it is defined by $\|x\| = \sqrt{\langle x, x \rangle}$.

Any orthogonal system of nonzero vectors can be normalized. If S is an orthogonal system, then the collection

$$S' = \left\{ \frac{x}{\|x\|} : x \in S \right\}$$

is an orthonormal system.

Theorem 2.30. *Every orthogonal system in an inner product space is linearly independent.*

Proof. Let S be an orthogonal system in an inner product space $(X, \langle\cdot,\cdot\rangle)$ over $\mathbb{K}$. Assume that $\sum_{i=1}^{n} \alpha_i x_i = 0$ for some $x_1, x_2, \ldots, x_n \in S$ and $\alpha_1, \alpha_2, \ldots, \alpha_n \in \mathbb{K}$. Then we have

$$\begin{aligned} 0 &= \sum_{i=1}^{n} \langle 0, \alpha_i x_i \rangle = \sum_{i=1}^{n} \left\langle \sum_{j=1}^{n} \alpha_j x_j, \alpha_i x_i \right\rangle = \sum_{i=1}^{n} \langle \alpha_i x_i, \alpha_i x_i \rangle \\ &= \sum_{i=1}^{n} |\alpha_i|^2 \|x_i\|^2, \end{aligned}$$

which implies that $\alpha_i = 0$ for all $i \in \{1, 2, \ldots, n\}$. Therefore, $x_1, x_2, \ldots, x_n$ are linearly independent. □

Definition 2.31. Any sequence of vectors in an inner product space is called an *orthonormal sequence* if they form an orthonormal system.

For each $i \in \mathbb{N}$, we define $e_i = (0, \ldots, 0, 1, 0, \ldots)$ with 1 in the ith position. Then the set $S = \{e_1, e_2, \ldots\}$ is an orthonormal sequence in l^2. We recall that l^2 is the space of all sequences $\{x_i\}_{i \in \mathbb{N}}$ of complex numbers such that $\sum_{i=1}^{\infty} |x_i|^2 < \infty$ with the inner product defined by

$$\langle x, y \rangle = \sum_{i=1}^{\infty} x_i \overline{y}_i.$$

According to Theorem 2.27, the Pythagorean theorem holds for every pair of orthogonal vectors in an inner product space.

In the following theorem, we will generalize the Pythagorean theorem to a finite number of orthogonal vectors.

Theorem 2.32 (Extended Pythagorean Theorem). *If $(X, \langle\cdot,\cdot\rangle)$ is an inner product space, then*

$$\left\| \sum_{i=1}^{n} x_i \right\|^2 = \sum_{i=1}^{n} \|x_i\|^2$$

for all orthogonal vectors $x_1, x_2, \ldots, x_n \in X$.

Proof. We will use mathematical induction. By Theorem 2.27, $\|x_1 + x_2\|^2 = \|x_1\|^2 + \|x_2\|^2$ for any pair of orthogonal vectors x_1 and x_2. Hence, our assertion is true for $n = 2$. Now we assume that our assertion is true for $n - 1$, i.e.,

$$\left\|\sum_{i=1}^{n-1} x_i\right\|^2 = \sum_{i=1}^{n-1} \|x_i\|^2$$

for all orthogonal vectors $x_1, x_2, \ldots, x_{n-1} \in X$.

Let $x_1, x_2, \ldots, x_n$ be orthogonal vectors in X. We set $x = \sum_{i=1}^{n-1} x_i$ and $y = x_n$. Then $x \perp y$. Thus we have

$$\left\|\sum_{i=1}^{n} x_i\right\|^2 = \|x + y\|^2 = \|x\|^2 + \|y\|^2 = \sum_{i=1}^{n-1} \|x_i\|^2 + \|x_n\|^2 = \sum_{i=1}^{n} \|x_i\|^2,$$

which proves our assertion. $\square$

The Bessel inequality gives us information about the coefficients of an element in an inner product space with respect to an orthonormal set.

Theorem 2.33 (Bessel Inequality). *Given $n \in \mathbb{N}$, assume that $\{x_1, x_2, \ldots, x_n\}$ is an orthonormal set of vectors in an inner product space $(X, \langle\cdot,\cdot\rangle)$. Then*

$$\left\|x - \sum_{i=1}^{n} \langle x, x_i\rangle x_i\right\|^2 = \|x\|^2 - \sum_{i=1}^{n} |\langle x, x_i\rangle|^2$$

and

$$\sum_{i=1}^{n} |\langle x, x_i\rangle|^2 \leq \|x\|^2$$

for all $x \in X$.

Proof. By Theorem 2.32, we have

$$\left\|\sum_{i=1}^{n} \alpha_i x_i\right\|^2 = \sum_{i=1}^{n} \|\alpha_i x_i\|^2 = \sum_{i=1}^{n} |\alpha_i|^2$$

for all scalars $\alpha_1, \alpha_2, \ldots, \alpha_n$. Hence, we get

$$\begin{aligned}
&\left\| x - \sum_{i=1}^{n} \alpha_i x_i \right\|^2 \\
&= \left\langle x - \sum_{i=1}^{n} \alpha_i x_i, \, x - \sum_{j=1}^{n} \alpha_j x_j \right\rangle \\
&= \|x\|^2 - \left\langle x, \sum_{j=1}^{n} \alpha_j x_j \right\rangle - \left\langle \sum_{i=1}^{n} \alpha_i x_i, x \right\rangle + \sum_{i=1}^{n} \sum_{j=1}^{n} \alpha_i \overline{\alpha}_j \langle x_i, x_j \rangle \\
&= \|x\|^2 - \sum_{j=1}^{n} \overline{\alpha}_j \langle x, x_j \rangle - \sum_{i=1}^{n} \alpha_i \overline{\langle x, x_i \rangle} + \sum_{i=1}^{n} \alpha_i \overline{\alpha}_i \\
&= \|x\|^2 - \sum_{i=1}^{n} |\langle x, x_i \rangle|^2 + \sum_{i=1}^{n} |\langle x, x_i \rangle - \alpha_i|^2.
\end{aligned}$$

Putting $\alpha_i = \langle x, x_i \rangle$ yields the Bessel equality. Finally, the Bessel inequality is an immediate consequence of the Bessel equality. □

Remark 2.34. Let $\{x_i\}_{i \in \mathbb{N}}$ be an orthonormal sequence of vectors in an inner product space $(X, \langle \cdot, \cdot \rangle)$.

(i) If we let $n \to \infty$ in the Bessel equality, then we get

$$\sum_{i=1}^{\infty} |\langle x, x_i \rangle|^2 \leq \|x\|^2$$

for any $x \in X$, which implies that the series $\sum_{i=1}^{\infty} |\langle x, x_i \rangle|^2$ converges for any $x \in X$. In other words, the sequence $\{\langle x, x_i \rangle\}_{i \in \mathbb{N}}$ is an element of l^2.

(ii) The expansion

$$x \sim \sum_{i=1}^{\infty} \langle x, x_i \rangle x_i$$

is called a *generalized Fourier series* of x. The scalar $\langle x, x_i \rangle$ is called the *generalized Fourier coefficient* of x with respect to the orthonormal sequence $\{x_i\}_{i \in \mathbb{N}}$. In general, we do not know whether the generalized Fourier series of x converges.

The following theorem states that the generalized Fourier series of x converges in any Hilbert space.

Theorem 2.35. *Let $\{x_i\}_{i\in\mathbb{N}}$ be an orthonormal sequence in a Hilbert space $(X, \langle\cdot,\cdot\rangle)$ and let $\{\alpha_i\}_{i\in\mathbb{N}}$ be a sequence of scalars. Then the series $\sum_{i=1}^{\infty} \alpha_i x_i$ converges if and only if $\sum_{i=1}^{\infty} |\alpha_i|^2 < \infty$. In that case, it holds that*

$$\left\| \sum_{i=1}^{\infty} \alpha_i x_i \right\|^2 = \sum_{i=1}^{\infty} |\alpha_i|^2.$$

Proof. By Theorem 2.32, it holds that

$$\left\| \sum_{i=m}^{n} \alpha_i x_i \right\|^2 = \sum_{i=m}^{n} |\alpha_i|^2 \tag{2.6}$$

for all integers m and n with $n > m > 0$. We define $s_n = \sum_{i=1}^{n} \alpha_i x_i$ for each $n \in \mathbb{N}$. If $\sum_{i=1}^{\infty} |\alpha_i|^2 < \infty$, then it follows from (2.6) that $\{s_n\}_{n\in\mathbb{N}}$ is a Cauchy sequence in X. Since X is complete, the series $\sum_{i=1}^{\infty} \alpha_i x_i$ converges.

Now we define $t_n = \sum_{i=1}^{n} |\alpha_i|^2$ for any $n \in \mathbb{N}$. If the series $\sum_{i=1}^{\infty} \alpha_i x_i$ converges, then it follows from (2.6) that $\sum_{i=1}^{\infty} |\alpha_i|^2$ converges because $\{t_n\}_{n\in\mathbb{N}}$ is a Cauchy sequence in $\mathbb{R}$.

Putting $m = 1$ and letting $n \to \infty$ in (2.6), we complete our proof. □

By Remark 2.34 (i) and Theorem 2.35, it holds that in any Hilbert space X the series $\sum_{i=1}^{\infty} \langle x, x_i\rangle x_i$ converges for each $x \in X$. However, it should be noted that it may converge to an element different from x.

Now we shall treat separately the inner product spaces in which the generalized Fourier series of x necessarily converges to x.

Definition 2.36. Assume that $(X, \langle\cdot,\cdot\rangle)$ is an inner product space. An orthonormal sequence $\{x_i\}_{i\in\mathbb{N}}$ in X is called *complete* if

$$x = \sum_{i=1}^{\infty} \langle x, x_i\rangle x_i$$

or equivalently

$$\lim_{n\to\infty} \left\| x - \sum_{i=1}^{n} \langle x, x_i\rangle x_i \right\| = 0$$

for all $x \in X$.

An orthonormal set that forms a basis is called an orthonormal basis.

Definition 2.37. Assume that $(X, \langle \cdot, \cdot \rangle)$ is an inner product space. An orthonormal system $\mathcal{B}$ in X is called an *orthonormal basis* if each element x of X has a unique representation

$$x = \sum_{i=1}^{\infty} \alpha_i x_i,$$

where the α_i's are scalars and x_i's are distinct elements of $\mathcal{B}$.

Remark 2.38. Let $(X, \langle \cdot, \cdot \rangle)$ be a Hilbert space over $\mathbb{K}$.

(i) A complete orthonormal sequence $\{x_i\}_{i \in \mathbb{N}}$ in X is an orthonormal basis in X.

(ii) If $\{x_i\}_{i \in \mathbb{N}}$ is a complete orthonormal sequence in X, then the space spanned by $\{x_i\}_{i \in \mathbb{N}}$

$$\operatorname{span}(x_1, x_2, \ldots) = \left\{ \sum_{i=1}^{n} \alpha_i x_i : n \in \mathbb{N};\ \alpha_1, \alpha_2, \ldots, \alpha_n \in \mathbb{K} \right\}$$

is dense in X.

Proof. (i) It suffices to prove the uniqueness of the generalized Fourier series of x. If there are two generalized Fourier series of x

$$x = \sum_{i=1}^{\infty} \alpha_i x_i \quad \text{and} \quad x = \sum_{i=1}^{\infty} \beta_i x_i,$$

then it follows from Theorem 2.35 that

$$0 = \|x - x\|^2 = \left\| \sum_{i=1}^{\infty} \alpha_i x_i - \sum_{i=1}^{\infty} \beta_i x_i \right\|^2 = \left\| \sum_{i=1}^{\infty} (\alpha_i - \beta_i) x_i \right\|^2 = \sum_{i=1}^{\infty} |\alpha_i - \beta_i|^2,$$

which implies that $\alpha_i = \beta_i$ for every $i \in \mathbb{N}$.

(ii) We can easily prove this claim using Theorem 1.10. We encourage the reader to try this proof as an exercise. □

Some important characterizations of complete orthonormal sequences in Hilbert spaces are given in the following theorems.

Theorem 2.39. *Let $(X, \langle\cdot,\cdot\rangle)$ be a Hilbert space and let x be an arbitrary element of X. An orthonormal sequence $\{x_i\}_{i\in\mathbb{N}}$ in X is complete if and only if $\langle x, x_i\rangle = 0$ for all $i \in \mathbb{N}$ implies $x = 0$.*

Proof. Assume that $\{x_i\}_{i\in\mathbb{N}}$ is a complete orthonormal sequence in X. Then, each element x of X has the representation

$$x = \sum_{i=1}^{\infty} \langle x, x_i\rangle x_i.$$

Hence, if $\langle x, x_i\rangle = 0$ for all $i \in \mathbb{N}$, then $x = 0$.

Conversely, assume that $\langle x, x_i\rangle = 0$ for all $i \in \mathbb{N}$ implies $x = 0$. For any $x \in X$, let

$$y = \sum_{i=1}^{\infty} \langle x, x_i\rangle x_i.$$

By Remark 2.34 (i) and Theorem 2.35, the sum y exists in X. Thus, we have

$$\begin{aligned}\langle x - y, x_n\rangle &= \langle x, x_n\rangle - \langle y, x_n\rangle \\ &= \langle x, x_n\rangle - \left\langle \sum_{i=1}^{\infty} \langle x, x_i\rangle x_i,\, x_n \right\rangle \\ &= \langle x, x_n\rangle - \sum_{i=1}^{\infty} \langle x, x_i\rangle \langle x_i, x_n\rangle,\end{aligned}$$

and since $\{x_i\}_{i\in\mathbb{N}}$ is an orthonormal sequence in X, we obtain

$$\langle x - y, x_n\rangle = \langle x, x_n\rangle - \langle x, x_n\rangle = 0$$

for all $n \in \mathbb{N}$. Thus, it follows from the assumption that $x - y = 0$. Therefore, $x = y = \sum_{i=1}^{\infty} \langle x, x_i\rangle x_i$ for any $x \in X$. According to Definition 2.36, the orthonormal sequence $\{x_i\}_{i\in\mathbb{N}}$ is complete. □

We remind here once again that the norm in the inner product space means the norm derived from the inner product.

Theorem 2.40 (Parseval's Formula). *Let $(X, \langle\cdot,\cdot\rangle)$ be a Hilbert space and let $\{x_i\}_{i\in\mathbb{N}}$ be an orthonormal sequence in X. Then $\{x_i\}_{i\in\mathbb{N}}$ is complete if and only if*

$$\|x\|^2 = \sum_{i=1}^{\infty} |\langle x, x_i\rangle|^2$$

for all $x \in X$.

Proof. Let x be an arbitrary element of X. By Theorem 2.33, we have

$$\left\| x - \sum_{i=1}^{n} \langle x, x_i \rangle x_i \right\|^2 = \|x\|^2 - \sum_{i=1}^{n} |\langle x, x_i \rangle|^2 \tag{2.7}$$

for any $n \in \mathbb{N}$. If $\{x_i\}_{i\in\mathbb{N}}$ is a complete orthonormal sequence in X, then

$$\lim_{n\to\infty} \left(\|x\|^2 - \sum_{i=1}^{n} |\langle x, x_i \rangle|^2 \right) = 0,$$

i.e.,

$$\|x\|^2 = \sum_{i=1}^{\infty} |\langle x, x_i \rangle|^2$$

for all $x \in X$.

Conversely, assume that the last equality holds for all $x \in X$. Then the expression on the right of (2.7) converges to 0 as $n \to \infty$. Hence, it follows from (2.7) that

$$\lim_{n\to\infty} \left\| x - \sum_{i=1}^{n} \langle x, x_i \rangle x_i \right\|^2 = 0$$

for all $x \in X$. Finally, it follows from Definition 2.36 that $\{x_i\}_{i\in\mathbb{N}}$ is complete. □

2.7 Orthogonal Complements

We remember that every inner product space is a normed space, so moreover it is a metric space.

A subspace S of a Hilbert space X is an inner product space. If we additionally assume that S is a closed subspace of X, then S is a complete inner product space by Theorem 1.41. That is, S is itself a Hilbert space.

In the following definition, we extend the concept of orthogonality defined in Definition 2.26.

Definition 2.41. Let S be a nonempty subset of a Hilbert space $(X, \langle\cdot,\cdot\rangle)$. An element x of X is said to be *orthogonal* to S if $\langle x, y \rangle = 0$ for all $y \in S$. In this case, we write $x \perp S$. The set of all elements of X which are orthogonal to S, denoted by $S^\perp$, is said to be the *orthogonal complement* of S.

If an element x of a Hilbert space X satisfies the condition $x \perp y$ for all $y \in X$, then $x = 0$. Hence, $X^\perp = \{0\}$. Similarly, $\{0\}^\perp = X$.

Theorem 2.42. *Let S be a subset of a Hilbert space X. Then the orthogonal complement $S^\perp$ of S is a closed subspace of X.*

Proof. For all scalars α, β and $x, y \in S^\perp$, it holds that

$$\langle \alpha x + \beta y, z\rangle = \alpha\langle x, z\rangle + \beta\langle y, z\rangle = 0$$

for all $z \in S$, i.e., $\alpha x + \beta y \in S^\perp$ for all scalars α, β and $x, y \in S^\perp$. Thus, $S^\perp$ is a vector subspace of X.

Let $\{x_i\}_{i\in\mathbb{N}}$ be an arbitrary sequence in $S^\perp$ that converges to some element x of X. According to Theorem 2.24, $\langle\cdot, y\rangle$ is a continuous function. Hence, we get

$$\langle x, y\rangle = \left\langle \lim_{i\to\infty} x_i, y\right\rangle = \lim_{i\to\infty}\langle x_i, y\rangle = 0$$

for any $y \in S$. That is, $x \in S^\perp$. This fact implies that $S^\perp$ is closed. □

Theorem 2.42 states that the orthogonal complement $S^\perp$ is a Hilbert space for every subset S of a Hilbert space.

Definition 2.43. A subset S of a vector space is called *convex* if $\alpha x+(1-\alpha)y \in S$ for all $x, y \in S$ and all real numbers α with $0 < \alpha < 1$.

We note that every vector subspace is a convex set. We now introduce a theorem about the minimization of the norm, which is of fundamental importance in approximation theory.

Theorem 2.44. *Let X be a Hilbert space. If S is a closed convex subset of X and x is an arbitrary element of X, then there exists a unique element y_x of S such that*

$$\|x - y_x\| = \inf_{z\in S}\|x - z\|.$$

Proof. Assume that $\{y_i\}_{i\in\mathbb{N}}$ is a sequence in S that satisfies

$$\lim_{i\to\infty}\|x - y_i\| = \inf_{z\in S}\|x - z\|.$$

Since $\frac{1}{2}(y_i + y_j) \in S$, we have

$$\left\|x - \frac{1}{2}(y_i + y_j)\right\| \geq \inf_{z\in S}\|x - z\|$$

for all $i, j \in \mathbb{N}$.

Furthermore, it follows from Theorem 2.23 that

$$\begin{aligned}\|y_i - y_j\|^2 &= 4\left\|x - \frac{1}{2}(y_i + y_j)\right\|^2 + \|y_i - y_j\|^2 - 4\left\|x - \frac{1}{2}(y_i + y_j)\right\|^2 \\ &= \|(x - y_i) + (x - y_j)\|^2 + \|(x - y_i) - (x - y_j)\|^2 \\ &\quad - 4\left\|x - \frac{1}{2}(y_i + y_j)\right\|^2 \\ &= 2\big(\|x - y_i\|^2 + \|x - y_j\|^2\big) - 4\left\|x - \frac{1}{2}(y_i + y_j)\right\|^2.\end{aligned}$$

From the facts

$$2\big(\|x - y_i\|^2 + \|x - y_j\|^2\big) \to 4\Big(\inf_{z\in S} \|x - z\|\Big)^2,$$

as $i, j \to \infty$, and

$$-4\left\|x - \frac{1}{2}(y_i + y_j)\right\|^2 \le -4\Big(\inf_{z\in S} \|x - z\|\Big)^2,$$

it follows that $\|y_i - y_j\|^2 \to 0$ as $i, j \to \infty$. Hence, $\{y_i\}_{i\in\mathbb{N}}$ is a Cauchy sequence in S.

Since S is complete as a closed subset of a complete space X by Theorem 1.41, the limit $\lim_{i\to\infty} y_i = y_x$ exists and $y_x \in S$. Moreover, it follows from the continuity of norm (Theorem 2.24) that

$$\|x - y_x\| = \left\|x - \lim_{i\to\infty} y_i\right\| = \lim_{i\to\infty} \|x - y_i\| = \inf_{z\in S} \|x - z\|.$$

We now want to prove the uniqueness of y_x. Assume that there exists another element y'_x of S that satisfies

$$\|x - y'_x\| = \inf_{z\in S} \|x - z\|.$$

By the parallelogram law (Theorem 2.23), we have

$$\begin{aligned}\|y_x - y'_x\|^2 &= \|(x - y'_x) - (x - y_x)\|^2 \\ &= 2\|x - y'_x\|^2 + 2\|x - y_x\|^2 - \|(x - y'_x) + (x - y_x)\|^2 \\ &= 4\Big(\inf_{z\in S} \|x - z\|\Big)^2 - \|2x - (y_x + y'_x)\|^2.\end{aligned}$$

Since $\frac{1}{2}(y_x + y'_x) \in S$, we obtain

$$\|y_x - y'_x\|^2 = 4\Big(\inf_{z\in S} \|x - z\|\Big)^2 - 4\left\|x - \frac{1}{2}(y_x + y'_x)\right\|^2 \le 0,$$

which implies that $y_x = y'_x$. □

The following theorem states that every Hilbert space is the direct sum of a closed subspace and its orthogonal complement.

We denote by $\Re\alpha$ and $\Im\alpha$ the real part and the imaginary part of any complex number α.

Theorem 2.45. *Let X be a Hilbert space over $\mathbb{K}$. If S is a closed subspace of X, then each element x of X has a unique decomposition in the form $x = y + z$, where $y \in S$ and $z \in S^\perp$.*

Proof. We note that $S^\perp$ is a closed subspace of X by Theorem 2.42. If $x \in S$, then $x = x + 0$ is a unique decomposition, where $x \in S$ and $0 \in S^\perp$.

Assume now that $x \notin S$. Since S is a closed convex subset of X, it follows from Theorem 2.44 that there exists a unique element y_x of S such that

$$\|x - y_x\| = \inf_{z \in S} \|x - z\|.$$

Now we prove that $x = y_x + (x - y_x)$ is the decomposition that satisfies $y_x \in S$ and $x - y_x \in S^\perp$. If $w \in S$ and $\alpha \in \mathbb{K}$, then $y_x + \alpha w \in S$ and

$$\|x - y_x\|^2 \leq \|x - (y_x + \alpha w)\|^2 = \|x - y_x\|^2 - 2\Re\alpha\langle w, x - y_x\rangle + |\alpha|^2\|w\|^2.$$

Hence, we have

$$-2\Re\alpha\langle w, x - y_x\rangle + |\alpha|^2\|w\|^2 \geq 0. \tag{2.8}$$

We assume that α is a positive real number. Then we divide inequality (2.8) by α and let $\alpha \to 0$ to get

$$\Re\langle w, x - y_x\rangle \leq 0. \tag{2.9}$$

Similarly, we replace α in inequality (2.8) by $-i\alpha$ ($\alpha > 0$) and divide the resulting inequality by α, and we let $\alpha \to 0$ to obtain

$$\Im\langle w, x - y_x\rangle \leq 0. \tag{2.10}$$

Since S is a vector space, $w \in S$ implies $-w \in S$. Hence, we can replace w with $-w$ in inequalities (2.9) and (2.10). Thus, it follows that $\Re\langle w, x - y_x\rangle = \Im\langle w, x - y_x\rangle = 0$, i.e., $\langle w, x - y_x\rangle = 0$ for any $w \in S$, which implies that $x - y_x \in S^\perp$.

Finally, we prove the uniqueness of decomposition $x = y + z$, where $y \in S$ and $z \in S^\perp$. We note that if $x = y' + z'$, $y' \in S$, and $z' \in S^\perp$, then $y - y' \in S$ and $z' - z \in S^\perp$. Since $y - y' = z' - z$, it should hold that $y - y' = z' - z = 0$. □

Remark 2.46. Let S be a closed subspace of a Hilbert space X.

(i) Each element of X can be uniquely expressed as the sum of an element of S and an element of $S^\perp$. We express this symbolically simply as

$$X = S \oplus S^\perp$$

and we say that X is the *direct sum* of S and $S^\perp$. Moreover, the above equality is called an *orthogonal decomposition* of X.

(ii) The union of a basis of S and a basis of $S^\perp$ is a basis of X.

2.8 Separable Hilbert Spaces

As we see in the following definition, a Hilbert space is said to be separable if it has a countable orthonormal basis.

Definition 2.47. A Hilbert space X is called *separable* if there exists a complete orthonormal sequence in X.

We recall that l^2 is the space of all sequences $\{x_i\}_{i\in\mathbb{N}}$ of complex (or real) numbers such that $\sum_{i=1}^{\infty} |x_i|^2 < \infty$. For any $i \in \mathbb{N}$, we set $e_i = (0, \ldots, 0, 1, 0, \ldots)$ with 1 in the ith position. Then we can easily prove that the orthonormal sequence $\{e_1, e_2, \ldots\}$ in l^2 is complete. Hence l^2 is a separable Hilbert space endowed with the inner product given by

$$\langle x, y\rangle = \sum_{i=1}^{\infty} x_i \overline{y}_i$$

for all elements $x = \{x_i\}_{i\in\mathbb{N}}$ and $y = \{y_i\}_{i\in\mathbb{N}}$ of l^2.

In view of Definition 1.9, a metric space is said to be separable if it contains a countable dense subset. For this reason, the following theorem is often used as a definition of separability for the Hilbert space.

Theorem 2.48. *Each separable Hilbert space contains a countable dense subset.*

Proof. Let X be a separable Hilbert space. Since X is a separable Hilbert space, there exists a complete orthonormal sequence $\{x_i\}_{i\in\mathbb{N}}$ in X. We define

$$\begin{aligned} S = \big\{&(\alpha_1 + i\beta_1)x_1 + \cdots + (\alpha_n + i\beta_n)x_n : \\ &\alpha_1, \ldots, \alpha_n, \beta_1, \ldots, \beta_n \in \mathbb{Q};\ n \in \mathbb{N}\big\}, \end{aligned}$$

where $\mathbb{Q}$ denotes the set of all rational numbers. Then S is countable. We note that for all $x \in X$,

$$\left\| \sum_{i=1}^{n} \langle x, x_i \rangle x_i - x \right\| \to 0 \quad \text{as} \quad n \to \infty.$$

Since $\sum_{i=1}^{n} \langle x, x_i \rangle x_i \in S$, S is dense in X. □

Theorem 2.49. *Each orthogonal set in a separable Hilbert space is countable.*

Proof. Assume that S is an arbitrary orthogonal set in a separable Hilbert space X and S' is the set of normalized vectors from S, i.e.,

$$S' = \left\{ \frac{1}{\|z\|} z : z \in S \right\}.$$

Since S' is an orthonormal set, we have

$$\begin{aligned} \|x - y\|^2 &= \langle x - y, x - y \rangle \\ &= \langle x, x \rangle - \langle x, y \rangle - \langle y, x \rangle + \langle y, y \rangle \\ &= 2 \end{aligned}$$

for all distinct $x, y \in S'$, which implies that the distance between any two distinct elements of S' is $\sqrt{2}$.

We now consider the collection of all open balls with radius $\frac{1}{\sqrt{2}}$ and center at every points in S'. It is obvious that no two of these open balls can have a common point. Since every dense subset of X has at least one element in each open ball and X has a countable dense subset, S' has to be countable. Therefore, S is countable, which completes the proof. □

Let X_1 and X_2 be vector spaces over $\mathbb{K}$. A mapping $L : X_1 \to X_2$ is called a *linear mapping* if $L(\alpha x + \beta y) = \alpha L(x) + \beta L(y)$ for all $x, y \in X_1$ and $\alpha, \beta \in \mathbb{K}$.

Definition 2.50. Let X_1 and X_2 be Hilbert spaces over the same scalar field. X_1 is called *isomorphic* to X_2 if there exists a one-to-one linear mapping T from X_1 onto X_2 such that

$$\langle T(x), T(y) \rangle = \langle x, y \rangle$$

for all $x, y \in X_1$. In the case, the mapping T is called a *Hilbert space isomorphism* of X_1 onto X_2.

According to the following theorem, any separable infinite-dimensional Hilbert space is isometric to the space l^2 of all square-summable sequences. On the other hand, a separable n-dimensional Hilbert space is isometric to either $\mathbb{C}^n$ or $\mathbb{R}^n$, depending on whether the scalar field $\mathbb{K}$ is either $\mathbb{C}$ or $\mathbb{R}$.

Theorem 2.51. *Let X be a separable Hilbert space over $\mathbb{K}$ and let n be a positive integer.*

(i) *If X is infinite-dimensional, then it is isomorphic to l^2.*

(ii) *If X is n-dimensional, then it is isomorphic to $\mathbb{K}^n$.*

Proof. Since X is a separable Hilbert space, there exists a complete orthonormal sequence $\{x_i\}_{i\in\mathbb{N}}$ in X.

(i) First we consider the case where X is infinite-dimensional. Then it follows that $\{x_i\}_{i\in\mathbb{N}}$ is an infinite sequence. We define a mapping T by

$$T(x) = \big(\langle x, x_1\rangle, \langle x, x_2\rangle, \ldots, \langle x, x_i\rangle, \ldots\big)$$

for all $x \in X$. Then, due to Theorem 2.35, T is a one-to-one mapping from X onto l^2. In addition, T is a linear mapping.

Furthermore, considering the definition of inner product on l^2 and continuity of inner product, we have

$$\begin{aligned}
\langle T(x), T(y)\rangle &= \langle(\langle x, x_1\rangle, \langle x, x_2\rangle, \ldots), (\langle y, x_1\rangle, \langle y, x_2\rangle, \ldots)\rangle \\
&= \sum_{i=1}^{\infty} \langle x, x_i\rangle\overline{\langle y, x_i\rangle} \\
&= \sum_{i=1}^{\infty} \big\langle x, \langle y, x_i\rangle x_i\big\rangle \\
&= \left\langle x, \sum_{i=1}^{\infty} \langle y, x_i\rangle x_i\right\rangle \\
&= \langle x, y\rangle
\end{aligned}$$

for all $x, y \in X$. Therefore, we conclude that T is a Hilbert space isomorphism from X onto l^2.

(ii) In the same way, we can prove the second claim. The proof of this is left to the reader. □

We can show that the isomorphism of Hilbert spaces is an equivalence relation. In a sense, there is only one real and one complex infinite-dimensional separable Hilbert space.

Chapter 3

Measure Theory

The concept of a measure is a generalization and formalization of length, area, volume, and other common notions such as mass and probability of events. These seemingly different concepts have many similarities and can often be unified by the concept of measure. Measures are fundamental to probability theory and integration theory. In this chapter, we briefly present the basic concepts and theorems of general measure theory that are essential for the proof of Ulam's conjecture, mainly with reference to C. A. Rogers' book [18].

3.1 Outer Measures

We remember that the power set of a set X is the collection of all subsets of X and we denote the power set of X with $\mathcal{P}(X)$. In measure theory, an outer measure is a function defined on the power set of a given set with values in the extended real numbers that satisfy some additional conditions.

Definition 3.1. Let $\mathcal{P}(X)$ be the power set of a set X. A function $\mu^* : \mathcal{P}(X) \to [0, \infty]$ is called an *outer measure* on X if it satisfies the following conditions:

(i) $\mu^*(\emptyset) = 0$.

(ii) If $A_1, A_2 \in \mathcal{P}(X)$ with $A_1 \subset A_2$, then $\mu^*(A_1) \leq \mu^*(A_2)$.

(iii) If $\{A_i\}_{i\in\mathbb{N}}$ is a sequence in $\mathcal{P}(X)$, then

$$\mu^*\left(\bigcup_{i=1}^{\infty} A_i\right) \leq \sum_{i=1}^{\infty} \mu^*(A_i).$$

S.-M. Jung, *Ulam's Conjecture on Invariance of Measure in the Hilbert Cube*, Frontiers in Mathematics, https://doi.org/10.1007/978-3-031-30886-4_3

For example, if $\mu^*(A)$ is the Lebesgue outer measure for any subset A of the n-dimensional Euclidean space $\mathbb{R}^n$, then μ^* is an outer measure on $\mathbb{R}^n$.

Although outer measures are defined on the power set of the space, the class of all measurable sets defined below has special and useful properties.

Definition 3.2. If μ^* is an outer measure on a set X, a subset E of X is called μ^**-measurable* if

$$\mu^*(A \cup B) = \mu^*(A) + \mu^*(B)$$

for all subsets A and B of X with $A \subset E$ and $B \subset X \setminus E$.

Remark 3.3. Let A, B, and E be subsets of a set X and let μ^* be an outer measure on X.

(i) A and B are said to be *separated* by E if $A \subset E$ and $B \subset X \setminus E$.

(ii) E is said to be μ^*-measurable if the outer measure μ^* is additive on sets that are separated by E.

Before proving the theorem that provides considerable information about the measurable sets, it is convenient to introduce the concept of σ-algebras and to prove a simple lemma on σ-algebras.

Definition 3.4. A collection $\mathcal{A}$ of subsets of a set X is said to be a σ*-algebra* if it satisfies the following conditions:

(i) $\emptyset \in \mathcal{A}$.

(ii) If $A \in \mathcal{A}$, then $X \setminus A \in \mathcal{A}$.

(iii) If $A_i \in \mathcal{A}$ for all $i \in \mathbb{N}$, then $\bigcup_{i=1}^{\infty} A_i \in \mathcal{A}$.

It follows from (ii) and (iii) that if $A_i \in \mathcal{A}$ for all $i \in \mathbb{N}$, then

$$\bigcap_{i=1}^{\infty} A_i = X \setminus \left(\bigcup_{i=1}^{\infty} (X \setminus A_i) \right) \in \mathcal{A},$$

which implies that every σ-algebra is closed under the countable intersection.

Now we prove a simple lemma giving some necessary and sufficient conditions specifying σ-algebras.

Lemma 3.5. *If a collection $\mathcal{A}$ of subsets of a set X satisfies the following conditions:*

(*i*) $\emptyset \in \mathcal{A}$.

(*ii*) *If* $A \in \mathcal{A}$, *then* $X \setminus A \in \mathcal{A}$.

(*iii*) *If* $A_1, A_2 \in \mathcal{A}$, *then* $A_1 \cup A_2 \in \mathcal{A}$.

(*iv*) *If* $A_i \in \mathcal{A}$ *for all* $i \in \mathbb{N}$ *and they are disjoint, then* $\bigcup_{i=1}^{\infty} A_i \in \mathcal{A}$,

then $\mathcal{A}$ *is a* σ*-algebra.*

Proof. Let $\{A_i\}_{i\in\mathbb{N}}$ be an arbitrary sequence in $\mathcal{A}$. It follows from (i) and (iii) that

$$\bigcup_{j=1}^{i-1} A_j \in \mathcal{A}$$

for any $i \in \mathbb{N}$, where we set $\bigcup_{j=1}^{0} A_j = \emptyset$. Furthermore, by (ii), we have

$$X \setminus \bigcup_{j=1}^{i-1} A_j \in \mathcal{A}$$

for all $i \in \mathbb{N}$. Moreover, it follows from (ii) and (iii) that

$$A_i \cap \left(X \setminus \bigcup_{j=1}^{i-1} A_j \right) = X \setminus \left((X \setminus A_i) \cup \left(\bigcup_{j=1}^{i-1} A_j \right) \right) \in \mathcal{A}$$

for any $i \in \mathbb{N}$. Since the above sets are disjoint sets of $\mathcal{A}$, by (iv), we get

$$\bigcup_{i=1}^{\infty} A_i = \bigcup_{i=1}^{\infty} \left(A_i \cap \left(X \setminus \bigcup_{j=1}^{i-1} A_j \right) \right) \in \mathcal{A},$$

which implies that $\mathcal{A}$ is a σ-algebra. □

We can now prove a theorem that gives considerable information about the measurable sets and the behavior of the outer measure on the measurable sets.

Theorem 3.6. *Let* μ^* *be an outer measure on a set* X.

(*i*) *If* $\mu^*(N) = 0$, *then* N *is* μ^**-measurable.*

(*ii*) *If* A *is* μ^**-measurable, so is* $X \setminus A$.

(*iii*) *If* $\{A_i\}_{i\in\mathbb{N}}$ *is a sequence of* μ^**-measurable sets, then* $\bigcup_{i=1}^{\infty} A_i$ *and* $\bigcap_{i=1}^{\infty} A_i$ *are* μ^**-measurable.*

(*iv*) *If* $\{A_i\}_{i\in\mathbb{N}}$ *is a disjoint sequence of* μ^**-measurable sets, then*

$$\mu^*\left(\bigcup_{i=1}^{\infty} A_i\right) = \sum_{i=1}^{\infty} \mu^*(A_i).$$

Proof. (a) Assume that N is a subset of X with $\mu^*(N) = 0$ and that A and B are subsets of X with $A \subset N$ and $B \subset X \setminus N$. Then, by Definition 3.1, we have

$$\mu^*(B) \leq \mu^*(A \cup B) \leq \mu^*(A) + \mu^*(B) \leq \mu^*(N) + \mu^*(B) = \mu^*(B).$$

Hence, it follows that

$$\mu^*(A \cup B) = \mu^*(A) + \mu^*(B),$$

which implies that N is μ^*-measurable, i.e., the statement in (i) is true.

(b) Assume that A is μ^*-measurable. Then

$$\mu^*(B \cup C) = \mu^*(B) + \mu^*(C)$$

for all subsets B and C of X with $B \subset A$ and $C \subset X \setminus A$. Since $B \subset A$ and $C \subset X \setminus A$ if and only if $C \subset X \setminus A$ and $B \subset X \setminus (X \setminus A) = A$, we conclude that $X \setminus A$ is μ^*-measurable, i.e., the statement (ii) is true.

(c) Let A_1 and A_2 be μ^*-measurable sets. Assume that B and C are arbitrary subsets of X with $\mu^*(B) < \infty$, $\mu^*(C) < \infty$ and with

$$B \subset A_1 \cup A_2 \quad \text{and} \quad C \subset X \setminus (A_1 \cup A_2).$$

We note that $B \cup C = (B \cap A_1) \cup ((B \cup C) \cap (X \setminus A_1))$ and the sets $B \cap A_1$ and $(B \cup C) \cap (X \setminus A_1)$ are separated by the μ^*-measurable set A_1. Thus, we get

$$\mu^*(B \cup C) = \mu^*(B \cap A_1) + \mu^*\big((B \cup C) \cap (X \setminus A_1)\big). \tag{3.1}$$

On the other hand, we have

$$(B \cup C) \cap (X \setminus A_1) = \big(B \cap (X \setminus A_1)\big) \cup C$$

and the sets $B \cap (X \setminus A_1)$ and C are separated by the μ^*-measurable set A_2. Hence, we obtain

$$\mu^*\big((B \cup C) \cap (X \setminus A_1)\big) = \mu^*\big(B \cap (X \setminus A_1)\big) + \mu^*(C). \tag{3.2}$$

Finally, since A_1 is a μ^*-measurable set, we have

$$\mu^*(B \cap A_1) + \mu^*(B \cap (X \setminus A_1)) = \mu^*(B). \tag{3.3}$$

Therefore, it follows from (3.1), (3.2), and (3.3) that

$$\mu^*(B \cup C) = \mu^*(B) + \mu^*(C).$$

If one of the subsets B and C of X has an infinite μ^*-outer measure, then the above equality is obviously true, which implies that $A_1 \cup A_2$ is μ^*-measurable. We have thus proved (iii) for the finite sequence of μ^*-measurable sets.

(d) Let $\{A_i\}_{i \in \mathbb{N}}$ be a disjoint sequence of μ^*-measurable sets. We set

$$A = \bigcup_{i=1}^{\infty} A_i$$

and let B and C be arbitrary subsets of X that satisfy the conditions

$$B \subset A \quad \text{and} \quad C \subset X \setminus A.$$

By applying mathematical induction to (c), we see that the set $\bigcup\limits_{i=1}^{n} A_i$ is μ^*-measurable for any $n \in \mathbb{N}$. Since $C \subset X \setminus A \subset X \setminus \bigcup\limits_{i=1}^{n} A_i$, we have

$$\mu^*(B \cup C) \geq \mu^*\left(\left(B \cap \bigcup_{i=1}^{n} A_i\right) \cup C\right) = \mu^*\left(B \cap \bigcup_{i=1}^{n} A_i\right) + \mu^*(C). \tag{3.4}$$

Since the sets $A_1, A_2, \ldots, A_n$ are disjoint and μ^*-measurable, we get

$$\begin{aligned}
\mu^*\left(B \cap \bigcup_{i=1}^{n} A_i\right) &= \mu^*\left(\left(B \cap \bigcup_{i=1}^{n-1} A_i\right) \cup (B \cap A_n)\right) \\
&= \mu^*\left(B \cap \bigcup_{i=1}^{n-1} A_i\right) + \mu^*(B \cap A_n) \\
&= \mu^*\left(B \cap \bigcup_{i=1}^{n-2} A_i\right) + \mu^*(B \cap A_{n-1}) + \mu^*(B \cap A_n) \\
&= \cdots \\
&= \mu^*(B \cap A_1) + \mu^*(B \cap A_2) + \cdots + \mu^*(B \cap A_n).
\end{aligned}$$

Then, it follows from (3.4) and the last equality that

$$\mu^*(B \cup C) \geq \sum_{i=1}^{n} \mu^*(B \cap A_i) + \mu^*(C)$$

for all $n \in \mathbb{N}$, which implies that

$$\mu^*(B \cup C) \geq \sum_{i=1}^{\infty} \mu^*(B \cap A_i) + \mu^*(C).$$

Thus, due to Definition 3.1 (iii), we obtain

$$\begin{aligned}
\mu^*(B \cup C) &\geq \sum_{i=1}^{\infty} \mu^*(B \cap A_i) + \mu^*(C) \\
&\geq \mu^*\left(B \cap \bigcup_{i=1}^{\infty} A_i\right) + \mu^*(C) \qquad (3.5) \\
&= \mu^*(B) + \mu^*(C) \\
&\geq \mu^*(B \cup C).
\end{aligned}$$

That is, $\mu^*(B \cup C) = \mu^*(B) + \mu^*(C)$ for all subsets B and C of X with $B \subset A$ and $C \subset X \setminus A$, where $A = \bigcup_{i=1}^{\infty} A_i$. In other words, $\bigcup_{i=1}^{\infty} A_i$ is μ^*-measurable.

On the other hand, assuming $C = \emptyset$ in (3.5), we get

$$\mu^*(B) = \sum_{i=1}^{\infty} \mu^*(B \cap A_i) \tag{3.6}$$

for every subset B of $A = \bigcup_{i=1}^{\infty} A_i$. Moreover, taking $B = A$ in the last equality, we obtain

$$\mu^*\left(\bigcup_{i=1}^{\infty} A_i\right) = \sum_{i=1}^{\infty} \mu^*(A_i).$$

Therefore, we have proved (iv) and also (iii) for the disjoint sequence of μ^*-measurable sets.

(e) According to (a), the empty set $\emptyset$ is μ^*-measurable. Moreover, using (a), (b), (c), and (d), together with Lemma 3.5, we conclude that the collection of all μ^*-measurable sets in X is a σ-algebra. Hence, the collection of all μ^*-measurable subsets of X is closed under the operations of countable union and intersection. This completes the proof of this theorem. □

Corollary 3.7. *Let μ^* be an outer measure on a set X. Assume that E is an arbitrary subset of X and $\{A_i\}_{i \in \mathbb{N}}$ is a disjoint sequence of μ^*-measurable subsets of X. Then*

$$\mu^*\left(E \cap \bigcup_{i=1}^{\infty} A_i\right) = \sum_{i=1}^{\infty} \mu^*(E \cap A_i).$$

Proof. Keeping in mind that $\{A_i\}_{i\in\mathbb{N}}$ is a disjoint sequence, we put $B = E \cap \bigcup_{i=1}^{\infty} A_i$ in (3.6) to obtain

$$\mu^*\left(E \cap \bigcup_{i=1}^{\infty} A_i\right) = \sum_{i=1}^{\infty} \mu^*\left(E \cap \left(\bigcup_{j=1}^{\infty} A_j\right) \cap A_i\right) = \sum_{i=1}^{\infty} \mu^*(E \cap A_i),$$

which completes the proof. □

3.2 Measures in Abstract Spaces

At this point, it is convenient to introduce the concept of a measure defined on a σ-algebra of sets, sometimes called a countably additive measure by some other mathematicians.

Definition 3.8. Let X be a set and let $\mathcal{A}$ be a σ-algebra of subsets of X. A set function $\mu : \mathcal{A} \to [0, \infty]$ is called a *measure* if it satisfies the conditions:

(i) $\mu(\emptyset) = 0$.

(ii) For any disjoint sequence $\{A_i\}_{i\in\mathbb{N}}$ in $\mathcal{A}$, it holds that

$$\mu\left(\bigcup_{i=1}^{\infty} A_i\right) = \sum_{i=1}^{\infty} \mu(A_i).$$

In view of this definition, we can rephrase Theorem 3.6 as follows:

Theorem 3.9. *Let X be a set and let μ^* be an outer measure on X.*

(i) *The collection $\mathcal{M}$ of all μ^*-measurable subsets of X is a σ-algebra containing the null sets, where a subset N of X is called a null set if $\mu^*(N) = 0$.*

(ii) *The restriction of μ^* to $\mathcal{M}$ is a measure on $\mathcal{M}$.*

It is desirable to have a general method for constructing an outer measure from a pre-measure defined on any class of sets including $\emptyset$. Following M. E. Munroe [15], we will call this method *Method I.*

Definition 3.10. Given a set X, let $\mathcal{C}$ be some collection of subsets of X. A set function $\tau : \mathcal{C} \to [0, \infty]$ is called a *pre-measure* if it satisfies the following conditions:

(i) $\emptyset \in \mathcal{C}$.

(ii) $\tau(\emptyset) = 0$.

The assumption in Definition 3.10 that $\emptyset$ belongs to $\mathcal{C}$ is useful in that finite unions can appear as special cases of infinite unions in the following theorem. Here we introduce Method I by M. E. Munroe (see [15, Theorem 11.3]).

Theorem 3.11 (Munroe). *Let τ be a pre-measure defined on a collection $\mathcal{C}$ of subsets of a set X. If the set function $\mu^* : \mathcal{P}(X) \to [0, \infty]$ is defined by*

$$\mu^*(E) = \inf\left\{ \sum_{i=1}^{\infty} \tau(C_i) : E \subset \bigcup_{i=1}^{\infty} C_i \textit{ where } C_i \in \mathcal{C} \textit{ for all } i \in \mathbb{N} \right\}$$

for all $E \in \mathcal{P}(X)$, then μ^* is an outer measure on X.

Proof. (a) Since $0 \leq \tau(C) \leq \infty$ for any $C \in \mathcal{C}$, it holds that

$$0 \leq \mu^*(E) \leq \infty$$

for all subsets E of X.

(b) It holds that

$$\begin{aligned} \mu^*(\emptyset) &= \inf\left\{ \sum_{i=1}^{\infty} \tau(C_i) : \emptyset \subset \bigcup_{i=1}^{\infty} C_i \text{ where } C_i \in \mathcal{C} \text{ for all } i \in \mathbb{N} \right\} \\ &\leq \sum_{i=1}^{\infty} \tau(\emptyset) \\ &= 0. \end{aligned}$$

Thus, we have $\mu^*(\emptyset) = 0$.

(c) If $E_1 \subset E_2$, then every covering of E_2 is also a covering of E_1 and thus

$$\mu^*(E_1) \leq \mu^*(E_2).$$

(d) Assume that $\{E_i\}_{i \in \mathbb{N}}$ is an arbitrary sequence of subsets of X. We assert that

$$\mu^*\left(\bigcup_{i=1}^{\infty} E_i \right) \leq \sum_{i=1}^{\infty} \mu^*(E_i).$$

Our assertion is obviously true when $\sum\limits_{i=1}^{\infty} \mu^*(E_i) = \infty$. Hence, we assume that $\sum\limits_{i=1}^{\infty} \mu^*(E_i) < \infty$. Then, $\mu^*(E_i) < \infty$ for any $i \in \mathbb{N}$.

Given any $\varepsilon > 0$ and integer $i > 1$, there exists a sequence $\{C_{ij}\}_{j\in\mathbb{N}}$ in $\mathcal{C}$ such that

$$E_i \subset \bigcup_{j=1}^{\infty} C_{ij} \quad \text{and} \quad \sum_{j=1}^{\infty} \tau(C_{ij}) \leq \mu^*(E_i) + \frac{1}{2^i}\varepsilon.$$

Assume that $\{D_i\}_{i\in\mathbb{N}}$ is a sequence obtained by rearranging the double sequence $\{C_{ij}\}_{i,j\in\mathbb{N}}$ as a single sequence. Then, we see that

$$\bigcup_{i=1}^{\infty} E_i \subset \bigcup_{i=1}^{\infty} D_i \quad \text{and} \quad D_i \in \mathcal{C} \quad (\text{for all } i \in \mathbb{N})$$

and

$$\begin{aligned}
\mu^*\left(\bigcup_{i=1}^{\infty} E_i\right) \leq \sum_{i=1}^{\infty} \tau(D_i) = \sum_{i=1}^{\infty}\sum_{j=1}^{\infty} \tau(C_{ij}) &\leq \sum_{i=1}^{\infty}\left(\mu^*(E_i) + \frac{1}{2^i}\varepsilon\right) \\
&= \sum_{i=1}^{\infty} \mu^*(E_i) + \varepsilon,
\end{aligned}$$

which proves our assertion, since we can choose ε arbitrarily small. □

Now we will show that any outer measure μ^* defined on X can be obtained by applying Method I to an appropriate pre-measure.

Theorem 3.12. *Let X be a set. Every outer measure μ^* on X can be constructed, by Method I, from the pre-measure $\tau : \mathcal{P}(X) \to [0, \infty]$ that satisfies $\tau(A) = \mu^*(A)$ for all subsets A of X.*

Proof. Since $\emptyset \in \mathcal{P}(X)$ and $\tau(\emptyset) = \mu^*(\emptyset) = 0$, τ is a pre-measure. Based on Theorem 3.11, we can construct an outer measure from τ by applying Method I and we will call this outer measure λ^*. Then, for any subset A of X, we have

$$\begin{aligned}
\lambda^*(A) &= \inf\left\{\sum_{i=1}^{\infty} \tau(C_i) : A \subset \bigcup_{i=1}^{\infty} C_i \text{ where } C_i \in \mathcal{P}(X) \text{ for all } i \in \mathbb{N}\right\} \\
&\leq \tau(A) \\
&= \mu^*(A),
\end{aligned}$$

since A belongs to $\mathcal{P}(X)$ and covers A.

On the other hand, if $A \subset \bigcup_{i=1}^{\infty} C_i$ and $C_i \in \mathcal{P}(X)$ for all i, then

$$\mu^*(A) \leq \mu^*\left(\bigcup_{i=1}^{\infty} C_i\right) \leq \sum_{i=1}^{\infty} \mu^*(C_i) = \sum_{i=1}^{\infty} \tau(C_i),$$

which implies that

$$\lambda^*(A) = \inf\left\{\sum_{i=1}^{\infty}\tau(C_i) : A \subset \bigcup_{i=1}^{\infty} C_i \text{ where } C_i \in \mathcal{P}(X) \text{ for all } i \in \mathbb{N}\right\}$$
$$\geq \mu^*(A).$$

Therefore, we conclude that $\lambda^*(A) = \mu^*(A)$ for all subsets A of X. □

We will now explore the relation between a given outer measure μ^* and the measure μ defined on the σ-algebra of all μ^*-measurable sets.

Theorem 3.13. *Let X be a set. Assume that μ is a measure defined on a σ-algebra $\mathcal{A}$ of subsets of X. Then μ is a pre-measure defined on $\mathcal{A}$. Let λ^* be the outer measure constructed from the pre-measure μ by Method I. Then the restriction of λ^* to the σ-algebra of all λ^*-measurable subsets of X is an extension of μ.*

Proof. It is obvious that μ is a pre-measure defined on $\mathcal{A}$. Thus, by Theorem 3.11, the set function λ^* constructed from μ by Method I is an outer measure on X.

Assume that E is any subset of X and $\{A_i\}_{i\in\mathbb{N}}$ is a sequence in $\mathcal{A}$ with

$$E \subset \bigcup_{i=1}^{\infty} A_i.$$

Then, since $\mathcal{A}$ is a σ-algebra, the sets

$$A = \bigcup_{i=1}^{\infty} A_i \quad \text{and} \quad B_i = A_i \setminus \bigcup_{j=1}^{i-1} A_j \quad (i \in \mathbb{N})$$

belong to the σ-algebra $\mathcal{A}$. Since μ is a measure on $\mathcal{A}$, we have

$$\mu(A) = \mu\left(\bigcup_{i=1}^{\infty} B_i\right) = \sum_{i=1}^{\infty}\mu(B_i) \leq \sum_{i=1}^{\infty}\mu(A_i).$$

Therefore, we get

$$\lambda^*(E) = \inf\left\{\sum_{i=1}^{\infty}\mu(A_i) : E \subset \bigcup_{i=1}^{\infty} A_i \text{ where } A_i \in \mathcal{A} \text{ for all } i \in \mathbb{N}\right\}$$
$$\geq \inf\left\{\mu(A) : E \subset A \text{ and } A \in \mathcal{A}\right\}.$$

On the other hand, taking any set A in $\mathcal{A}$ with $E \subset A$ and considering the covering of E by the sequence $\{A, \emptyset, \emptyset, \ldots\}$, we have

$$\lambda^*(E) \leq \inf\left\{\mu(A) : E \subset A \text{ and } A \in \mathcal{A}\right\},$$

and thus, we conclude that

$$\lambda^*(E) = \inf\{\mu(A) : E \subset A \text{ and } A \in \mathcal{A}\}. \tag{3.7}$$

We can choose a sequence $\{A_i\}_{i\in\mathbb{N}}$ in $\mathcal{A}$ such that

$$E \subset A_i \quad \text{and} \quad \lambda^*(E) \le \mu(A_i) \le \lambda^*(E) + \frac{1}{i}$$

for all $i \in \mathbb{N}$. Then, the set $A = \bigcap_{i=1}^{\infty} A_i$ belongs to $\mathcal{A}$ and satisfies

$$E \subset A \quad \text{and} \quad \lambda^*(E) = \mu(A), \tag{3.8}$$

and the infimum in (3.7) is calculated explicitly.

When E belongs to the σ-algebra $\mathcal{A}$, we have

$$\begin{aligned}\mu(E) &= \inf\{\mu(E) : E \subset A \text{ where } A \in \mathcal{A}\}\\ &\le \inf\{\mu(A) : E \subset A \text{ where } A \in \mathcal{A}\}\\ &\le \mu(E).\end{aligned}$$

Hence, it follows from (3.7) that

$$\lambda^*(E) = \mu(E)$$

for all $E \in \mathcal{A}$. Therefore, μ is the restriction of λ^* to the σ-algebra $\mathcal{A}$.

It remains to prove that each set of $\mathcal{A}$ is λ^*-measurable. Assume that A is an arbitrary set of $\mathcal{A}$ and that C and D are arbitrary subsets of X, which are separated by A, namely,

$$C \subset A \quad \text{and} \quad D \subset X \setminus A.$$

Then, for any $i \in \mathbb{N}$, we can select a set B_i from $\mathcal{A}$ such that

$$C \cup D \subset B_i \quad \text{and} \quad \lambda^*(C \cup D) \le \mu(B_i) \le \lambda^*(C \cup D) + \frac{1}{i}.$$

Since $C \subset A \cap B_i$ and $D \subset (X \setminus A) \cap B_i$, we get

$$\begin{aligned}\lambda^*(C) + \lambda^*(D) &\le \mu(A \cap B_i) + \mu((X \setminus A) \cap B_i)\\ &= \mu((A \cap B_i) \cup ((X \setminus A) \cap B_i))\\ &= \mu(B_i)\\ &\le \lambda^*(C \cup D) + \frac{1}{i}\end{aligned}$$

for all $i \in \mathbb{N}$, which implies that $\lambda^*(C) + \lambda^*(D) \leq \lambda^*(C \cup D)$. Therefore, each set A of $\mathcal{A}$ is λ^*-measurable. □

If in Theorem 3.13 the measure μ is the restriction of an outer measure μ^* to the collection of all μ^*-measurable sets, then the outer measure λ^* will generally not coincide with the original outer measure μ^*. But the two outer measures are equal if and only if μ^* is regular.

Definition 3.14. Let X be a set. An outer measure μ^* on X is called *regular* if for each subset E of X there exists a μ^*-measurable subset A of X such that

$$E \subset A \quad \text{and} \quad \mu^*(E) = \mu^*(A).$$

On the basis of this definition, we prove a corollary to Theorem 3.13.

Corollary 3.15. *The outer measure λ^* defined in Theorem 3.13 is regular.*

Proof. Since each set of $\mathcal{A}$ is λ^*-measurable, it follows from (3.8) that for any subset E of X, there exists a set $A \in \mathcal{A}$ such that

$$E \subset A \quad \text{and} \quad \lambda^*(E) = \mu(A) = \lambda^*(A),$$

which implies that the outer measure λ^* is regular. □

Theorem 3.16. *Let X be a set. Assume that μ^* is an outer measure on X and μ is the restriction of μ^* to the σ-algebra $\mathcal{M}$ of all μ^*-measurable subsets of X. Then μ is a pre-measure defined on $\mathcal{M}$, and the outer measure λ^* constructed from the pre-measure μ by Method I is regular. All μ^*-measurable sets are λ^*-measurable and all λ^*-measurable sets E with $\lambda^*(E) < \infty$ are μ^*-measurable. Moreover, λ^* coincides with μ^* if and only if μ^* is regular.*

Proof. Due to Theorem 3.9, the set function μ is a measure on the σ-algebra $\mathcal{M}$ of all μ^*-measurable sets. By Theorem 3.13, (3.8) and by Corollary 3.15, the outer measure λ^* is regular, coincides with μ^* and μ on $\mathcal{M}$, and satisfies

$$\begin{aligned} \lambda^*(E) &= \inf \{\mu(E') : E \subset E' \text{ where } E' \in \mathcal{M}\} \\ &= \inf \{\mu^*(E') : E \subset E' \text{ where } E' \in \mathcal{M}\} \end{aligned} \tag{3.9}$$

for all subsets E of X.

Moreover, due to Theorem 3.13, all μ^*-measurable sets are λ^*-measurable. Assume that E is any λ^*-measurable subset of X with $\lambda^*(E) < \infty$. According to (3.8), there exists a μ^*-measurable subset A of X such that

$$E \subset A \quad \text{and} \quad \lambda^*(E) = \mu(A) = \mu^*(A). \tag{3.10}$$

Hence, it follows from (3.10) that $\lambda^*(A) = \mu^*(A) = \lambda^*(E) < \infty$. Since E is λ^*-measurable, we have

$$\lambda^*(E) + \lambda^*(A \setminus E) = \lambda^*(A) = \lambda^*(E) < \infty.$$

Hence, we get $\lambda^*(A \setminus E) = 0$. Since λ^* coincides with μ^* on $\mathcal{M}$, we obtain $\mu^*(A \setminus E) = 0$. Therefore, the null set $N = A \setminus E$ is μ^*-measurable by Theorem 3.6. Since A is μ^*-measurable, it holds that $E = A \cap (X \setminus N)$ is μ^*-measurable.

Finally, since the outer measure λ^* is regular, λ^* and μ^* can only coincide if μ^* is regular. On the other hand, if μ^* is regular, then we have

$$\mu^*(E) = \inf \{\mu^*(A) : E \subset A \text{ where } A \in \mathcal{M}\}$$

for all subsets E of X. Therefore, in view of (3.9), λ^* coincides with μ^*. □

The following theorem gives two useful properties of measurable sets.

Theorem 3.17. *Let X be a set. Assume that μ^* is an outer measure on X and $\{A_i\}_{i \in \mathbb{N}}$ is a sequence of μ^*-measurable subsets of X.*

(*i*) *If $A_1 \subset A_2 \subset A_3 \subset \cdots$ and E is an arbitrary subset of X, then*

$$\mu^*\left(E \cap \bigcup_{i=1}^{\infty} A_i\right) = \sup \{\mu^*(E \cap A_i) : i \in \mathbb{N}\}.$$

(*ii*) *If $A_1 \supset A_2 \supset A_3 \supset \cdots$, E is an arbitrary subset of X, and $\mu^*(E \cap A_i) < \infty$ for some $i \in \mathbb{N}$, then*

$$\mu^*\left(E \cap \bigcap_{i=1}^{\infty} A_i\right) = \inf \{\mu^*(E \cap A_i) : i \in \mathbb{N}\}.$$

Proof. (i) We set $A_0 = \emptyset$ and $B_i = A_i \setminus A_{i-1}$ for all $i \in \mathbb{N}$. Then $\{B_i\}_{i \in \mathbb{N}}$ is a disjoint sequence of μ^*-measurable subsets of X. Due to Corollary 3.7, we have

$$\begin{aligned}
\mu^*\left(E \cap \bigcup_{i=1}^{\infty} A_i\right) &= \mu^*\left(E \cap \bigcup_{i=1}^{\infty} B_i\right) \\
&= \sum_{i=1}^{\infty} \mu^*(E \cap B_i) \\
&= \lim_{n\to\infty} \sum_{i=1}^{n} \mu^*(E \cap B_i) \\
&= \lim_{n\to\infty} \mu^*\left(E \cap \bigcup_{i=1}^{n} B_i\right) \\
&= \lim_{n\to\infty} \mu^*(E \cap A_n) \\
&= \sup\left\{\mu^*(E \cap A_i) : i \in \mathbb{N}\right\}.
\end{aligned}$$

(ii) Assume that $\mu^*(E \cap A_n) < \infty$ for some $n \in \mathbb{N}$. We set $B_i = A_n \setminus A_{n+i}$ for all $i \in \mathbb{N}$. Then $\{B_i\}_{i \in \mathbb{N}}$ is a sequence of μ^*-measurable subsets of X with

$$B_1 \subset B_2 \subset B_3 \subset \cdots .$$

It then follows from (i) that

$$\mu^*\left(E \cap \bigcup_{i=1}^{\infty} B_i\right) = \sup\left\{\mu^*(E \cap B_i) : i \in \mathbb{N}\right\}.$$

Thus, we have

$$\mu^*\left((E \cap A_n) \setminus \bigcap_{j=n+1}^{\infty} A_j\right) = \sup\left\{\mu^*(E \cap (A_n \setminus A_{n+i})) : i \in \mathbb{N}\right\}. \quad (3.11)$$

Since $\{A_i\}_{i \in \mathbb{N}}$ is a non-increasing sequence, it follows that

$$E \cap \bigcap_{j=n+1}^{\infty} A_j = (E \cap A_n) \cap \bigcap_{j=n+1}^{\infty} A_j, \quad E \cap A_{n+i} = (E \cap A_n) \cap A_{n+i}.$$

Moreover, we see that

$$\begin{aligned}
E \cap A_n &= \left((E \cap A_n) \cap \bigcap_{j=n+1}^{\infty} A_j\right) \cup \left((E \cap A_n) \setminus \bigcap_{j=n+1}^{\infty} A_j\right) \\
&= \big((E \cap A_n) \cap A_{n+i}\big) \cup \big((E \cap A_n) \setminus A_{n+i}\big).
\end{aligned}$$

Since each A_i is a μ^*-measurable subset of X, so is $\bigcap_{j=n+1}^{\infty} A_j$. Hence, it holds that

$$\begin{aligned}\mu^*(E \cap A_n) &= \mu^*\left(E \cap \bigcap_{j=n+1}^{\infty} A_j\right) + \mu^*\left((E \cap A_n) \setminus \bigcap_{j=n+1}^{\infty} A_j\right) \\ &= \mu^*(E \cap A_{n+i}) + \mu^*((E \cap A_n) \setminus A_{n+i})\end{aligned}$$

for all $i \in \mathbb{N}$. Since $\mu^*(E \cap A_n) < \infty$, so are the other measures in the above formulae. In addition, by (3.11), we get

$$\begin{aligned}&\mu^*(E \cap A_n) - \mu^*\left(E \cap \bigcap_{j=n+1}^{\infty} A_j\right) \\ &\quad = \sup\left\{\mu^*(E \cap A_n) - \mu^*(E \cap A_{n+i}) : i \in \mathbb{N}\right\}.\end{aligned}$$

Therefore, since $\{A_i\}_{i \in \mathbb{N}}$ is a non-increasing sequence, we obtain

$$\begin{aligned}\mu^*\left(E \cap \bigcap_{i=1}^{\infty} A_i\right) &= \mu^*\left(E \cap \bigcap_{j=n+1}^{\infty} A_j\right) \\ &= \inf\left\{\mu^*(E \cap A_{n+i}) : i \in \mathbb{N}\right\} \\ &= \inf\left\{\mu^*(E \cap A_i) : i \in \mathbb{N}\right\},\end{aligned}$$

which completes the proof. □

In the following theorem, we will give a method of obtaining a new outer measure from a given collection of outer measures.

Theorem 3.18. *Let X be a set and let Λ be an arbitrary index set. If μ_λ^* is an outer measure on X for all $\lambda \in \Lambda$, then*

$$\mu^*(E) = \sup\left\{\mu_\lambda^*(E) : \lambda \in \Lambda\right\} \quad \text{(for all subsets } E \text{ of } X)$$

is an outer measure on X.

Proof. It is easy to show that $0 \leq \mu^*(E) \leq \infty$ for all subsets E of X. Moreover, it holds that

$$\mu^*(\emptyset) = \sup\left\{\mu_\lambda^*(\emptyset) : \lambda \in \Lambda\right\} = 0.$$

Assume that E_1 and E_2 are arbitrary subsets of X with $E_1 \subset E_2$. Then we have

$$\mu^*(E_1) = \sup\left\{\mu_\lambda^*(E_1) : \lambda \in \Lambda\right\} \leq \sup\left\{\mu_\lambda^*(E_2) : \lambda \in \Lambda\right\} = \mu^*(E_2).$$

Finally, we assume that $\{E_i\}_{i\in\mathbb{N}}$ is an arbitrary sequence of subsets of X. Then, for all $\lambda \in \Lambda$, we have

$$\mu_\lambda^*\left(\bigcup_{i=1}^\infty E_i\right) \le \sum_{i=1}^\infty \mu_\lambda^*(E_i) \le \sum_{i=1}^\infty \mu^*(E_i).$$

Hence, we get

$$\mu^*\left(\bigcup_{i=1}^\infty E_i\right) \le \sum_{i=1}^\infty \mu^*(E_i),$$

which completes the proof. □

We now generalize the concept of the regular outer measures.

Definition 3.19. Let X be a set and let $\mathcal{R}$ be a collection of subsets of X. An outer measure μ^* on X is called $\mathcal{R}$-*regular* if for each subset E of X there exists a set R in $\mathcal{R}$ such that

$$E \subset R \quad \text{and} \quad \mu^*(E) = \mu^*(R).$$

Comparing this definition with Definition 3.14, we note that the outer measure μ^* is regular if μ^* is $\mathcal{M}$-regular, where $\mathcal{M}$ is the σ-algebra of all μ^*-measurable subsets of X.

Definition 3.20. Let X be a set and let $\mathcal{R}$ be a collection of subsets of X.

(i) We denote by $\mathcal{R}_\sigma$ the collection of all countable set unions, i.e.,

$$\mathcal{R}_\sigma = \left\{\bigcup_{i=1}^\infty R_i : R_i \in \mathcal{R} \text{ for all } i \in \mathbb{N}\right\}.$$

(ii) We denote by $\mathcal{R}_\delta$ the collection of all countable set intersections, i.e.,

$$\mathcal{R}_\delta = \left\{\bigcap_{i=1}^\infty R_i : R_i \in \mathcal{R} \text{ for all } i \in \mathbb{N}\right\}.$$

Theorem 3.21. *Let X be a set and let μ^* be the outer measure constructed by Method I from a pre-measure τ defined on a collection $\mathcal{C}$ of subsets of X with $X \in \mathcal{C}$. Then μ^* is $\mathcal{C}_{\sigma\delta}$-regular.*

Proof. Assume that E is an arbitrary subset of X. If $\mu^*(E) = \infty$, then we have

$$E \subset X \quad \text{and} \quad \mu^*(E) = \mu^*(X),$$

where $X \in \mathcal{C} \subset \mathcal{C}_{\sigma\delta}$. Hence, from now on, we only need to treat each subset E of X with $\mu^*(E) < \infty$.

Since

$$\mu^*(E) = \inf\left\{\sum_{i=1}^{\infty} \tau(C_i) : E \subset \bigcup_{i=1}^{\infty} C_i \text{ where } C_i \in \mathcal{C} \text{ for all } i \in \mathbb{N}\right\},$$

for any integer $j \in \mathbb{N}$, we can choose a sequence $\{C_{i,j}\}_{i\in\mathbb{N}}$ in $\mathcal{C}$ such that

$$E \subset \bigcup_{i=1}^{\infty} C_{i,j} \quad \text{and} \quad \sum_{i=1}^{\infty} \tau(C_{i,j}) < \mu^*(E) + \frac{1}{j}.$$

We write

$$D = \bigcap_{j=1}^{\infty} \bigcup_{i=1}^{\infty} C_{i,j}.$$

Then, it is obvious that $D \in \mathcal{C}_{\sigma\delta}$ and $E \subset D$. Furthermore,

$$\begin{aligned}
\mu^*(E) &\leq \mu^*(D) \\
&= \inf\left\{\sum_{i=1}^{\infty} \tau(C_i) : D \subset \bigcup_{i=1}^{\infty} C_i \text{ where } C_i \in \mathcal{C} \text{ for all } i \in \mathbb{N}\right\} \\
&\leq \sum_{i=1}^{\infty} \tau(C_{i,j}) \\
&< \mu^*(E) + \frac{1}{j}
\end{aligned}$$

for every $j \in \mathbb{N}$, since $\{C_{i,j}\}_{i\in\mathbb{N}}$ is a sequence (in $\mathcal{C}$) covering D. Thus, we have

$$\mu^*(E) = \mu^*(D).$$

Therefore, for any subset E of X, there exists a $D \in \mathcal{C}_{\sigma\delta}$ such that $E \subset D$ and $\mu^*(E) = \mu^*(D)$, which implies that the outer measure μ^* is $\mathcal{C}_{\sigma\delta}$-regular. □

Definition 3.22. Let X be a topological space. The *Borel σ-algebra* of subsets of X is the smallest σ-algebra containing all open (or closed) subsets of X. Each set in the Borel σ-algebra is called a *Borel set*.

Corollary 3.23. *Let X be a set and let μ^* be the outer measure constructed by Method I from a pre-measure τ.*

(i) *If the pre-measure τ is defined on the collection $\mathcal{G}$ of all open subsets of X, then μ^* is $\mathcal{G}_\delta$-regular.*

(ii) *If the pre-measure τ is defined on the Borel σ-algebra $\mathcal{B}$, then μ^* is $\mathcal{B}$-regular (Borel-regular).*

Proof. We note that $X \in \mathcal{G}$, $X \in \mathcal{B}$, and $\mathcal{G}_\sigma = \mathcal{G}$ and that the collection of Borel sets is closed under countable unions and under countable intersections. Then the assertions of this corollary are direct consequences of Theorem 3.21. □

3.3 Measures in Metric Spaces

In this section, we introduce a second method called Method II to get an outer measure from a pre-measure. The names Method I and Method II go back to Munroe [15].

Let (X, d) be a metric space. We recall that the *diameter* of a subset E of X is defined as

$$d(E) = \begin{cases} 0 & (\text{for } E = \emptyset), \\ \sup\{d(x,y) : x, y \in E\} & (\text{for a nonempty bounded set } E), \\ \infty & (\text{for an unbounded set } E). \end{cases}$$

Let $\mathcal{C}$ be a collection of subsets of X. For any $\delta > 0$, we define

$$\mathcal{C}_\delta = \{E \in \mathcal{C} : d(E) < \delta\}.$$

We call the outer measure μ^* defined in the following theorem the outer measure constructed from the pre-measure τ by *Method II*.

Theorem 3.24 (Munroe). *Let (X, d) be a metric space and let τ be a pre-measure defined on a collection $\mathcal{C}$ of subsets of X. If the set function $\mu^* : \mathcal{P}(X) \to [0, \infty]$ is defined by*

$$\mu^*(E) = \sup\{\mu^*_\delta(E) : \delta > 0\}, \tag{3.12}$$

where

$$\mu^*_\delta(E) = \inf\left\{\sum_{i=1}^{\infty} \tau(C_i) : E \subset \bigcup_{i=1}^{\infty} C_i \text{ where } C_i \in \mathcal{C}_\delta \text{ for all } i \in \mathbb{N}\right\} \tag{3.13}$$

for all $E \in \mathcal{P}(X)$, then μ^* is an outer measure on X.

Proof. It is obvious that the restriction τ_δ of τ to $\mathcal{C}_\delta$ is also a pre-measure for any $\delta > 0$. Then we have

$$\mu_\delta^*(E) = \inf\left\{\sum_{i=1}^{\infty} \tau_\delta(C_i) : E \subset \bigcup_{i=1}^{\infty} C_i \text{ where } C_i \in \mathcal{C}_\delta \text{ for all } i \in \mathbb{N}\right\}$$

for all subsets E of X. Hence, it follows from Theorem 3.11 that μ_δ^* is the outer measure constructed by Method I from the pre-measure τ_δ. By Theorem 3.18, the set function

$$\mu^*(E) = \sup\left\{\mu_\delta^*(E) : \delta > 0\right\} \quad \text{(for all subsets } E \text{ of } X)$$

is an outer measure. □

As δ gets smaller, the size of the class of coverings over which the infimum (3.13) is taken also reduces. Therefore, for any fixed subset E of X, $\mu_\delta^*(E)$ does not decrease when δ decreases, and it is the small values of δ that are relevant to the supremum (3.12). In fact, we could immediately replace the formula (3.12) with

$$\mu^*(E) = \lim_{\delta \to 0+} \mu_\delta^*(E). \tag{3.14}$$

Definition 3.25. Let A and B be disjoint nonempty subsets of a metric space (X, d). Then A and B are said to be *positively separated* if the distance

$$\inf\left\{d(x,y) : x \in A \text{ and } y \in B\right\}$$

separating A and B is positive.

The main advantages of the outer measures constructed by Method II over those constructed by Method I all stem from the following theorem, which shows that the outer measures by Method II are additive when considered on the union of a pair of positively separated sets.

Theorem 3.26. *Let μ^* be an outer measure on a metric space (X, d) constructed from a pre-measure τ by Method II. Then*

$$\mu^*(A \cup B) = \mu^*(A) + \mu^*(B)$$

for all disjoint nonempty subsets A and B of X that are positively separated.

Proof. Let A and B be disjoint nonempty subsets of X that are positively separated. Since μ^* is an outer measure on X, we have

$$\mu^*(A \cup B) \le \mu^*(A) + \mu^*(B).$$

So it suffices to prove that

$$\mu^*(A \cup B) \geq \mu^*(A) + \mu^*(B).$$

Without loss of generality, we may assume that $\mu^*(A \cup B) < \infty$.

Since A and B are disjoint nonempty subsets of X that are positively separated, we can choose a $\delta > 0$ such that

$$\inf \{d(x, y) : x \in A \text{ and } y \in B\} \geq \delta.$$

Let $\varepsilon > 0$ be given. Assume that δ_1, δ_2, and η are constants given by

$$0 < \delta_1 < d(X), \quad 0 < \delta_2 < d(X), \quad \text{and} \quad \eta = \min\left\{\delta_1, \delta_2, \frac{1}{2}\delta\right\}.$$

We see that $\mu^*(A \cup B) < \infty$ and

$$\mu^*(A \cup B) = \sup_{\varrho > 0} \inf \left\{ \sum_{i=1}^{\infty} \tau(C_i) : A \cup B \subset \bigcup_{i=1}^{\infty} C_i \text{ where } C_i \in \mathcal{C}_\varrho \text{ for all } i \right\},$$

where the pre-measure τ is assumed to be defined on a collection $\mathcal{C}$ of subsets of X. Thus, we have

$$\inf \left\{ \sum_{i=1}^{\infty} \tau(C_i) : A \cup B \subset \bigcup_{i=1}^{\infty} C_i \text{ where } C_i \in \mathcal{C}_\eta \text{ for all } i \in \mathbb{N} \right\} \leq \mu^*(A \cup B),$$

and we may choose a sequence $\{C_i\}_{i \in \mathbb{N}}$ from $\mathcal{C}_\eta$ such that

$$A \cup B \subset \bigcup_{i=1}^{\infty} C_i \quad \text{and} \quad \sum_{i=1}^{\infty} \tau(C_i) \leq \mu^*(A \cup B) + \varepsilon.$$

Now, we assume that there exists an $i \in \mathbb{N}$ such that

$$C_i \cap A \neq \emptyset \quad \text{and} \quad C_i \cap B \neq \emptyset. \tag{3.15}$$

Then there would exist a_0 and b_0 with

$$a_0 \in C_i \cap A \quad \text{and} \quad b_0 \in C_i \cap B$$

so that

$$d(a_0, b_0) \leq d(C_i) \leq \eta \leq \frac{1}{2}\delta \leq \frac{1}{2} \inf \{d(x, y) : x \in A \text{ and } y \in B\} \leq \frac{1}{2} d(a_0, b_0),$$

which would lead to a contradiction that $d(a_0, b_0) = 0$.

So, we conclude that there does not exist $i \in \mathbb{N}$ satisfying the conditions in (3.15). For each $i \in \mathbb{N}$, we set

$$A_i = \begin{cases} C_i & (\text{for } C_i \cap A \neq \emptyset), \\ \emptyset & (\text{for } C_i \cap A = \emptyset) \end{cases} \quad \text{and} \quad B_i = \begin{cases} C_i & (\text{for } C_i \cap B \neq \emptyset), \\ \emptyset & (\text{for } C_i \cap B = \emptyset). \end{cases}$$

Then we get

$$\bigcup_{i=1}^{\infty} A_i \supset \bigcup_{i=1}^{\infty} (C_i \cap A) = \left(\bigcup_{i=1}^{\infty} C_i \right) \cap A = A,$$
$$\bigcup_{i=1}^{\infty} B_i \supset \bigcup_{i=1}^{\infty} (C_i \cap B) = \left(\bigcup_{i=1}^{\infty} C_i \right) \cap B = B.$$

Moreover, $\{A_i\}_{i \in \mathbb{N}}$ and $\{B_i\}_{i \in \mathbb{N}}$ are sequences of sets from $\mathcal{C}_\delta$. Since the conditions of (3.15) are not met, it holds that either

$$\tau(A_i) + \tau(B_i) = \tau(\emptyset) + \tau(\emptyset) = 0$$

or

$$\tau(A_i) + \tau(B_i) = \tau(C_i) + \tau(\emptyset) = \tau(C_i)$$

for all $i \in \mathbb{N}$. In either case, we have

$$\tau(A_i) + \tau(B_i) \leq \tau(C_i)$$

for all $i \in \mathbb{N}$. Hence, we get

$$\sum_{i=1}^{\infty} \tau(A_i) + \sum_{i=1}^{\infty} \tau(B_i) \leq \sum_{i=1}^{\infty} \tau(C_i) \leq \mu^*(A \cup B) + \varepsilon.$$

Furthermore, it holds that

$$A_i \in \mathcal{C}, \quad d(A_i) < \eta \leq \delta_1, \quad A \subset \bigcup_{i=1}^{\infty} A_i,$$
$$B_i \in \mathcal{C}, \quad d(B_i) < \eta \leq \delta_2, \quad B \subset \bigcup_{i=1}^{\infty} B_i.$$

Hence, we obtain

$$\mu^*_{\delta_1}(A) \leq \sum_{i=1}^{\infty} \tau(A_i), \quad \mu^*_{\delta_2}(B) \leq \sum_{i=1}^{\infty} \tau(B_i),$$

and

$$\mu^*_{\delta_1}(A) + \mu^*_{\delta_2}(B) \le \mu^*(A \cup B) + \varepsilon$$

for all δ_1 and δ_2 with $0 < \delta_1 < d(X)$ and $0 < \delta_2 < d(X)$. Thus, we have

$$\mu^*(A) + \mu^*(B) \le \mu^*(A \cup B) + \varepsilon$$

for any $\varepsilon > 0$. Therefore, we conclude that

$$\mu^*(A) + \mu^*(B) \le \mu^*(A \cup B),$$

which completes the proof. □

3.4 Metric Outer Measures

The property of the outer measures constructed by Method II presented in Theorem 3.26 is so important that it can be used as a definition.

Definition 3.27. An outer measure μ^* on a metric space X is called a *metric outer measure* if

$$\mu^*(A \cup B) = \mu^*(A) + \mu^*(B)$$

for any disjoint nonempty subsets A and B of X that are positively separated.

According to Theorem 3.26, the outer measure on a metric space, constructed by Method II, is always a metric outer measure.

Theorem 3.28. *Let μ^* be a metric outer measure on a metric space (X, d). Assume that $\{A_i\}_{i\in\mathbb{N}}$ is a sequence of subsets of X which satisfies*

$$A_1 \subset A_2 \subset A_3 \subset \cdots \quad \text{and} \quad A = \bigcup_{i=1}^{\infty} A_i.$$

If the sets A_i and $A \setminus A_{i+1}$ are positively separated for each $i \in \mathbb{N}$, then

$$\mu^*\left(\bigcup_{i=1}^{\infty} A_i\right) = \sup\left\{\mu^*(A_i) : i \in \mathbb{N}\right\}.$$

Proof. It is obvious that

$$\mu^*(A) \ge \sup\left\{\mu^*(A_i) : i \in \mathbb{N}\right\}.$$

So it is enough to prove that

$$\mu^*(A) \leq \sup\left\{\mu^*(A_i) : i \in \mathbb{N}\right\}.$$

Without loss of generality, we assume that $\sup\left\{\mu^*(A_i) : i \in \mathbb{N}\right\} < \infty$.

For every integer $i > 1$, we set $D_1 = A_1$ and $D_i = A_i \setminus A_{i-1}$. If i and j are positive integers with $j > i + 1$, then we have

$$D_i \subset A_i \quad \text{and} \quad D_j \subset A \setminus A_{j-1} \subset A \setminus A_{i+1}.$$

Since A_i and $A \setminus A_{i+1}$ are positively separated, the sets D_i and D_j are positively separated whenever $j > i + 1$. Furthermore, the sets

$$\bigcup_{i=1}^{n} D_{2i+m} \quad \text{and} \quad D_{2n+m+2}$$

are positively separated for all $n \in \mathbb{N}$ and $m \in \{0, -1\}$. Since μ^* is a metric outer measure, we can apply Theorem 3.26 inductively to obtain

$$\mu^*\left(\bigcup_{i=1}^{n} D_{2i+m}\right) = \sum_{i=1}^{n} \mu^*(D_{2i+m})$$

for all $n \in \mathbb{N}$ and $m \in \{0, -1\}$.

Now we see that

$$A_{2n} = \left(\bigcup_{i=1}^{n} D_{2i}\right) \cup \left(\bigcup_{i=1}^{n} D_{2i-1}\right)$$

for all $n \in \mathbb{N}$. Hence, we have

$$\sum_{i=1}^{n} \mu^*(D_{2i+m}) = \mu^*\left(\bigcup_{i=1}^{n} D_{2i+m}\right) \leq \mu^*(A_{2n}) \leq \sup\left\{\mu^*(A_i) : i \in \mathbb{N}\right\} < \infty$$

for all $n \in \mathbb{N}$ and $m \in \{0, -1\}$. Hence, the series

$$\sum_{i=1}^{\infty} \mu^*(D_{2i}) \quad \text{and} \quad \sum_{i=1}^{\infty} \mu^*(D_{2i-1}) \tag{3.16}$$

converge.

Finally, it holds that

$$\begin{aligned}\mu^*(A) &= \mu^*\left(\bigcup_{i=1}^{\infty} A_i\right)\\ &= \mu^*\left(A_n \cup \bigcup_{i=n+1}^{\infty} D_i\right)\\ &\le \mu^*(A_n) + \sum_{i=n+1}^{\infty} \mu^*(D_i)\\ &\le \sup\left\{\mu^*(A_i) : i \in \mathbb{N}\right\} + \sum_{i=n+1}^{\infty} \mu^*(D_i)\end{aligned} \tag{3.17}$$

for all $n \in \mathbb{N}$. Since both series in (3.16) converge, if we let $n \to \infty$ in (3.17), then we get

$$\mu^*(A) \le \sup\left\{\mu^*(A_i) : i \in \mathbb{N}\right\},$$

which completes the proof. □

Theorem 3.29. *If μ^* is a metric outer measure on a metric space (X, d), then all closed subsets of X are μ^*-measurable.*

Proof. Assume that K is an arbitrary closed subset of X and that A and B are arbitrary subsets of X which are separated by K, i.e.,

$$A \subset K \quad \text{and} \quad B \subset X \setminus K.$$

Without loss of generality, we assume that both A and B are nonempty sets.

We claim that B is expressed as the union of an increasing sequence $\{B_i\}_{i \in \mathbb{N}}$ of subsets of X, where B_i and $B \setminus B_{i+1}$ are positively separated and A and B_i are positively separated for each $i \in \mathbb{N}$.

We define B_i by

$$B_i = \left\{x \in B : \inf_{y \in K} d(x, y) > \frac{1}{i}\right\}$$

for each $i \in \mathbb{N}$. Then we see that

$$B_1 \subset B_2 \subset B_3 \subset \cdots \subset B.$$

Since $B \subset X \setminus K$, if $b \in B$, then $b \notin K$ and b is not a limit point of K. Hence, b belongs to B_i for all sufficiently large integers i. Therefore, it holds that

$$B = \bigcup_{i=1}^{\infty} B_i.$$

Since $A \subset K$, A and B_i are positively separated for all $i \in \mathbb{N}$. Now we assume that b and c are arbitrary elements of X with

$$b \in B_i \quad \text{and} \quad c \in B \setminus B_{i+1} \tag{3.18}$$

for some $i \in \mathbb{N}$. Then, it follows from the definition of B_{i+1} that

$$\inf \big\{ d(c,y) : y \in K \big\} \leq \frac{1}{i+1}.$$

Hence, we can select an element c' of K such that

$$d(c,c') \leq \frac{1}{i+1/2}.$$

If we had

$$d(b,c) \leq \frac{1/2}{i(i+1/2)}$$

for some $i \in \mathbb{N}$, then we would have

$$\begin{aligned}
\inf \big\{ d(b,y) : y \in K \big\} &\leq d(b,c') \\
&\leq d(b,c) + d(c,c') \\
&\leq \frac{1/2}{i(i+1/2)} + \frac{1}{i+1/2} \\
&= \frac{1}{i}
\end{aligned}$$

for some $i \in \mathbb{N}$, a contradiction to the definition of B_i and the assumption $b \in B_i$. Therefore, for any pair b and c satisfying (3.18), it had to hold that

$$d(b,c) > \frac{1}{i(2i+1)}$$

and B_i and $B \setminus B_{i+1}$ are positively separated for any $i \in \mathbb{N}$.

Finally, we use the metric property of the outer measure and Theorem 3.28 to have

$$\begin{aligned}
\mu^*(A \cup B) &\geq \sup \big\{ \mu^*(A \cup B_i) : i \in \mathbb{N} \big\} \\
&= \sup \big\{ \mu^*(A) + \mu^*(B_i) : i \in \mathbb{N} \big\} \\
&= \mu^*(A) + \sup \big\{ \mu^*(B_i) : i \in \mathbb{N} \big\} \\
&= \mu^*(A) + \mu^*(B),
\end{aligned}$$

which implies that every closed subset K of X is μ^*-measurable. □

We recall that the Borel σ-algebra of subsets of a topological space X is the smallest σ-algebra containing all open (or closed) subsets of X and each set in the Borel σ-algebra is called a Borel set.

Theorem 3.30. *If μ^* is a metric outer measure on a metric space X, then all Borel sets in X are μ^*-measurable.*

Proof. Let $\mathcal{M}$ be the collection of all μ^*-measurable subsets of X. Then it follows from Theorems 3.9 and 3.29 that $\mathcal{M}$ is σ-algebra and $\mathcal{M}$ contains any closed subset of X. Therefore, in view of Definition 3.22, we conclude that $\mathcal{M}$ contains the Borel sets in X. □

We now examine the regularity of the outer measures constructed by Method II from a pre-measure.

Theorem 3.31. *Let μ^* be the outer measure on a metric space (X, d) constructed, by Method II, from a pre-measure τ defined on a collection $\mathcal{C}$ of subsets of X with $X \in \mathcal{C}$. Then μ^* is $\mathcal{C}_{\sigma\delta}$-regular.*

Proof. Given any $\delta > 0$, let μ_δ^* be the outer measure on X constructed, by Method I, from the restriction τ_δ of τ to the sets C of $\mathcal{C}$ with $d(C) < \delta$. Then it follows that

$$\mu^*(E) = \sup_{\delta>0} \mu_\delta^*(E)$$

for all subsets E of X.

Due to Theorem 3.21, every outer measure μ_δ^* is $\mathcal{C}_{\sigma\delta}$-regular. Hence, if E is an arbitrary subset of X, then we can choose a $\mathcal{C}_{\sigma\delta}$-set C_i (i.e., $C_i \in \mathcal{C}_{\sigma\delta}$) such that

$$E \subset C_i \quad \text{and} \quad \mu_{1/i}^*(E) = \mu_{1/i}^*(C_i)$$

for any $i \in \mathbb{N}$. Then, the set

$$C = \bigcap_{i=1}^{\infty} C_i$$

is a $\mathcal{C}_{\sigma\delta}$-set with $E \subset C$.

Moreover, for any $\delta > 0$, there exists an $i \in \mathbb{N}$ satisfying $0 < \frac{1}{i} < \delta$ and

$$\mu_\delta^*(C) \le \mu_{1/i}^*(C) \le \mu_{1/i}^*(C_i) = \mu_{1/i}^*(E) \le \mu^*(E)$$

for all $\delta > 0$, which implies that $\mu^*(C) \le \mu^*(E)$. Thus, we have

$$C \in \mathcal{C}_{\sigma\delta}, \quad E \subset C, \quad \mu^*(E) = \mu^*(C).$$

Therefore, μ^* is $\mathcal{C}_{\sigma\delta}$-regular. □

Corollary 3.32. *Let μ^* be the outer measure on a metric space X constructed, by Method II, from a pre-measure τ.*

(i) *If the pre-measure τ is defined on the collection $\mathcal{G}$ of all open subsets of X, then μ^* is $\mathcal{G}_\delta$-regular.*

(ii) *If the pre-measure τ is defined on the Borel σ-algebra $\mathcal{B}$, then μ^* is $\mathcal{B}$-regular (Borel-regular).*

In either case, the outer measure μ^* is regular.

Proof. We note that $X \in \mathcal{G}$, $X \in \mathcal{B}$, and $\mathcal{G}_\sigma = \mathcal{G}$ and that the collection of Borel sets is closed under countable unions and under countable intersections. Then the assertions of this corollary are direct consequences of Theorem 3.31. The last assertion is true by Theorems 3.26 and 3.30. □

3.5 Lebesgue Measures

In this section, we explore the special but most important Lebesgue measure using the general theory we have developed so far.

Definition 3.33. Given an $n \in \mathbb{N}$, let $\mathbb{R}^n$ be the n-dimensional Euclidean space.

(i) Let $a_1, a_2, \ldots, a_n$ and $b_1, b_2, \ldots, b_n$ be arbitrary real numbers with $a_i \leq b_i$ for all $i \in \{1, 2, \ldots, n\}$. The set

$$R(a,b) = \{(x_1, x_2, \ldots, x_n) \in \mathbb{R}^n : a_i < x_i < b_i \text{ for any } i \in \{1, 2, \ldots, n\}\}$$

is said to be the *open rectangle* with corners

$$a = (a_1, a_2, \ldots, a_n) \quad \text{and} \quad b = (b_1, b_2, \ldots, b_n),$$

where we set $R(a,b) = \emptyset$ if $a_i = b_i$ for some $i \in \{1, 2, \ldots, n\}$. We use the symbol $\mathcal{R}$ to denote the collection of all such open rectangles.

(ii) We define a set function $\tau : \mathcal{R} \to [0, \infty]$ by

$$\tau\big(R(a,b)\big) = \prod_{i=1}^{n} (b_i - a_i)$$

for all $R(a,b) \in \mathcal{R}$. Then $\tau(R(a,b))$ is the "elementary volume" of the open rectangle $R(a,b)$. Obviously, τ is a pre-measure defined on $\mathcal{R}$.

Remark 3.34. Even though we replace $R(a,b)$ in Definition 3.33 with the *half-open rectangle* or the *closed rectangle* with corners a and b defined by

$$R[a,b) = \{(x_1, x_2, \ldots, x_n) \in \mathbb{R}^n : a_i \leq x_i < b_i \text{ for any } i \in \{1, 2, \ldots, n\}\}$$

or

$$R[a,b] = \{(x_1, x_2, \ldots, x_n) \in \mathbb{R}^n : a_i \le x_i \le b_i \text{ for any } i \in \{1, 2, \ldots, n\}\},$$

where $R[a,b) = \emptyset$ if $a_i = b_i$ for some $i \in \{1, 2, \ldots, n\}$, and we define $\mathcal{R}$ to be the collection of all such half-open rectangles or closed rectangles, the following theorems about the properties of Lebesgue measures are still true.

Let μ^* and ν^* be the outer measures constructed from the pre-measure τ by Methods I and II, respectively. Then it is easy to verify that μ^* is the usual Lebesgue (outer) measure. Our experience so far makes us confident that the outer measure ν^* would have many advantages over the outer measure μ^*. In fact, however, μ^* shares all the advantages of ν^*, as the following theorem shows.

Theorem 3.35. *Assume that μ^* and ν^* are the outer measures constructed from the pre-measure τ by Methods I and II, respectively. Then $\mu^* = \nu^*$.*

Proof. By Theorems 3.11 and 3.24, it is obvious that $\mu^*(E) \le \nu^*(E)$ for all subsets E of $\mathbb{R}^n$. We now claim that $\nu^*(E) \le \mu^*(E)$.

Assume that E is an arbitrary subset of $\mathbb{R}^n$ with $\mu^*(E) < \infty$, without loss of generality. For any $\varepsilon > 0$, we can select a sequence $\{R_i\}_{i \in \mathbb{N}}$ in $\mathcal{R}$ such that

$$E \subset \bigcup_{i=1}^{\infty} R_i \quad \text{and} \quad \sum_{i=1}^{\infty} \tau(R_i) \le \mu^*(E) + \varepsilon. \tag{3.19}$$

Given any $\delta > 0$ and any $R(a,b) \in \mathcal{R}$, we can select a large integer N such that the N^n closed rectangles

$$a_i + \frac{r_i - 1}{N}(b_i - a_i) \le x_i \le a_i + \frac{r_i}{N}(b_i - a_i) \quad (\text{for } i \in \{1, 2, \ldots, n\}),$$

where $r_1, r_2, \ldots, r_n \in \{1, 2, \ldots, N\}$, and such that all the closed rectangles have diameter less than δ.

Then, for any sufficiently small $\eta > 0$, the N^n open rectangles $R(r_1, r_2, \ldots, r_n)$ given by

$$a_i + \frac{r_i - 1}{N}(b_i - a_i) - \eta < x_i < a_i + \frac{r_i}{N}(b_i - a_i) + \eta \quad (\text{for } i \in \{1, 2, \ldots, n\})$$

cover the open rectangle $R(a,b)$, where $d(R(r_1, r_2, \ldots, r_n)) < \delta$ for $r_1, r_2, \ldots, r_n \in \{1, 2, \ldots, N\}$, and have

$$\sum_{r_1=1}^{N} \sum_{r_2=1}^{N} \cdots \sum_{r_n=1}^{N} \tau\big(R(r_1, r_2, \ldots, r_n)\big) \le \tau\big(R(a,b)\big) + O(\eta),$$

where the O denotes Landau's symbol (O notation).

Applying this process to every open rectangle R_i, we replace R_i with a finite sub-collection $\{R_{ij}\}_{j\in\{1,2,\ldots,j(i)\}}$ of open rectangles from $\mathcal{R}$ of diameter less than δ that satisfies

$$R_i \subset \bigcup_{j=1}^{j(i)} R_{ij} \quad \text{and} \quad \sum_{j=1}^{j(i)} \tau(R_{ij}) \leq \tau(R_i) + \frac{1}{2^i}\varepsilon.$$

Then, by (3.19), the collection of all such open rectangles R_{ij} is a δ-covering of E with

$$\sum_{i=1}^{\infty}\sum_{j=1}^{j(i)} \tau(R_{ij}) \leq \mu^*(E) + 2\varepsilon.$$

Hence, we have

$$\inf\left\{ \sum_{i=1}^{\infty} \tau(R_i) : E \subset \bigcup_{i=1}^{\infty} R_i \text{ where } R_i \in \mathcal{R}_\delta \text{ for all } i \in \mathbb{N} \right\} \leq \mu^*(E) + 2\varepsilon,$$

which implies that $\nu^*(E) \leq \mu^*(E) + 2\varepsilon$. Therefore, since ε is an arbitrary positive real number, we conclude that $\nu^*(E) \leq \mu^*(E)$ for each subset E of $\mathbb{R}^n$. □

The following theorem seems obvious, but its proof is not so easy. In fact, this theorem was originally proved in the one-dimensional case by E. Borel [1].

Theorem 3.36. *Assume that μ^* is the outer measure constructed from the pre-measure τ by Method I. Then $\mu^*(R) = \tau(R)$ for all open rectangles R of $\mathcal{R}$.*

Proof. It follows from Theorem 3.11 that $\mu^*(R) \leq \tau(R)$ for all open rectangles R of $\mathcal{R}$. It remains to prove that $\mu^*(R) \geq \tau(R)$ for all open rectangles R of $\mathcal{R}$.

We consider a closed rectangle I of the form

$$I = [a_1, b_1] \times [a_2, b_2] \times \cdots \times [a_n, b_n],$$

where a_i and b_i are real numbers satisfying $a_i < b_i$ for each $i \in \{1, 2, \ldots, n\}$. Assume that $\{R_j\}_{j\in\mathbb{N}}$ is an arbitrary sequence of open rectangles from $\mathcal{R}$ with

$$I \subset \bigcup_{j=1}^{\infty} R_j.$$

We note that $\{R_j\}_{j\in\mathbb{N}}$ is an open covering of the compact subset I of $\mathbb{R}^n$. Hence, we can choose an $N \in \mathbb{N}$ such that

$$I \subset \bigcup_{j=1}^{N} R_j.$$

For any $j \in \{1, 2, \ldots, N\}$, let $I^\circ \cap R_j$ be the open rectangle of the form

$$I^\circ \cap R_j = (a_{1,j}, b_{1,j}) \times (a_{2,j}, b_{2,j}) \times \cdots \times (a_{n,j}, b_{n,j}).$$

For all $i \in \{1, 2, \ldots, n\}$, let $c_{i,1}, c_{i,2}, \ldots, c_{i,2N}$ ($c_{i,1} = a_i$ and $c_{i,2N} = b_i$) be a rearrangement of the numbers $a_{i,1}, b_{i,1}, a_{i,2}, b_{i,2}, \ldots, a_{i,N}, b_{i,N}$ in nondecreasing order.

For any finite sequence $\{k(1), k(2), \ldots, k(n)\}$ with $k(i) \in \{1, 2, \ldots, 2N-1\}$, we define the open rectangle $R(k(1), k(2), \ldots, k(n))$ by

$$\begin{aligned} &R(k(1), k(2), \ldots, k(n)) \\ &\quad = (c_{1,k(1)}, c_{1,k(1)+1}) \times (c_{2,k(2)}, c_{2,k(2)+1}) \times \cdots \times (c_{n,k(n)}, c_{n,k(n)+1}). \end{aligned}$$

When this open rectangle is nonempty, its midpoint is in I° and so is in one of the open rectangles $R_1, R_2, \ldots, R_N$. Hence, by the choice of the numbers $c_{i,k(i)}$, the whole rectangle $R(k(1), k(2), \ldots, k(n))$ lies in one of the rectangles $I^\circ \cap R_j$. However, by the choice of the numbers $c_{i,k(i)}$, the elementary volume of each open rectangle $I^\circ \cap R_j$ is the sum of the elementary volumes of the open rectangles $R(k(1), k(2), \ldots, k(n))$ that it contains. Thus we have

$$\begin{aligned} \sum_{j=1}^{\infty} \tau(R_j) &\geq \sum_{j=1}^{N} \tau(R_j) \\ &\geq \sum_{j=1}^{N} \tau(I^\circ \cap R_j) \\ &= \sum_{j=1}^{N} \sum_{R(k(1),k(2),\ldots,k(n)) \subset I^\circ \cap R_j} \prod_{i=1}^{n} \left(c_{i,k(i)+1} - c_{i,k(i)}\right) \\ &\geq \sum_{k(1)=1}^{2N-1} \sum_{k(2)=1}^{2N-1} \cdots \sum_{k(n)=1}^{2N-1} \prod_{i=1}^{n} \left(c_{i,k(i)+1} - c_{i,k(i)}\right) \\ &= \prod_{i=1}^{n} (c_{i,2N} - c_{i,1}) \\ &= \prod_{i=1}^{n} (b_i - a_i). \end{aligned}$$

Therefore, we get

$$\mu^*(I) \geq \prod_{i=1}^{n} (b_i - a_i).$$

Finally, for any given nonempty open rectangle I°, we can construct in I° closed rectangles with elementary volume, and hence with Lebesgue outer measure, as close as we please to $\tau(I^\circ)$. So we have $\mu^*(I^\circ) \geq \tau(I^\circ)$. Therefore, we conclude that $\mu^*(R) = \tau(R)$ for all open rectangles R of $\mathcal{R}$, which completes our proof. □

Definition 3.37. Let μ^* be an outer measure on a vector space X. For any subset E and element p of X, we define

$$E + p = \{x + p : x \in E\}.$$

The outer measure μ^* is called *translation-invariant* if

$$\mu^*(E + p) = \mu^*(E)$$

for all subsets E and elements p of X.

We note that the Lebesgue measure is the unique Borel-regular, translation-invariant outer measure on the Euclidean space $\mathbb{R}^n$, which assigns unit measure to the unit n-cube $[0, 1]^n$.

Let $(X_1, \mathcal{A}_1)$ and $(X_2, \mathcal{A}_2)$ be measurable spaces and let μ_1 and μ_2 be measures on them. We denote by $\mathcal{A}_1 \otimes \mathcal{A}_2$ the σ-algebra on the Cartesian product $X_1 \times X_2$ generated by subsets of the form

$$\{A_1 \times A_2 : A_1 \in \mathcal{A}_1 \text{ and } A_2 \in \mathcal{A}_2\}.$$

We define the *product measure* $\mu_1 \otimes \mu_2$ on the measurable space $(X_1 \times X_2, \mathcal{A}_1 \otimes \mathcal{A}_2)$ by $(\mu_1 \otimes \mu_2)(A_1 \times A_2) = \mu_1(A_1)\mu_2(A_2)$ for all $A_1 \in \mathcal{A}_1$ and $A_2 \in \mathcal{A}_2$.

The Lebesgue measure μ_n is a measure defined in an n-dimensional Euclidean space. Unfortunately, since the Hilbert cube is an infinite-dimensional space, it is not possible to define the Lebesgue measure in the Hilbert cube. Instead, in the Hilbert cube we can define the standard product probability measure that most closely resembles the Lebesgue measure.

Definition 3.38. If a measure π defined on the Borel σ-algebra of the Hilbert cube has the following properties, we call it the *standard product probability measure*:

(*i*) $\pi(I^\omega) = 1$.

(*ii*) $\pi(B) = \prod_{i=1}^{\infty} \mu_1(B_i)$ for all Borel subsets B of I^ω with $B = \prod_{i=1}^{\infty} B_i$, where μ_1 is the Lebesgue measure in $\mathbb{R}$ and each B_i is a Borel subset of $[0, 1]$.

3.6 Hausdorff Measures

The Hausdorff measure is a generalization of the traditional concepts of length, area, and volume to non-integer dimensions, such as the Hausdorff dimension of fractals. The Hausdorff measure is a kind of outer measure, named after F. Hausdorff, that assigns a number in $[0, \infty]$ to each subset of a metric space.

Definition 3.39. We denote by $\mathcal{H}$ the set of all functions $h : [0, \infty) \to [0, \infty]$ with the following four properties:

(i) $h(0) = 0$.

(ii) $h(t) > 0$ for any $t > 0$.

(iii) h is monotonically increasing.

(iv) h is continuous on the right for all $t \geq 0$.

Every function h in $\mathcal{H}$ is called a *Hausdorff function*.

It is well known that every Hausdorff function induces a corresponding Hausdorff measure.

Definition 3.40. Let (X, d) be a metric space and $\mathcal{T}(d)$ the topology for X generated by d, and let h be a Hausdorff function. We set $h(G) = h(d(G))$ for all $G \in \mathcal{T}(d)$. Then the outer measure constructed from the pre-measure h, defined on $\mathcal{T}(d)$, by Method II is called the *Hausdorff measure* corresponding to the Hausdorff function h and is denoted by μ^h.

Theorem 3.24 confirms that the Hausdorff measure μ^h is an outer measure on the metric space X as described in the definition above. Moreover, it immediately follows from Theorems 3.26 and 3.30 that all Borel sets are μ^h-measurable.

Theorem 3.41. *Let (X, d) be a metric space and let μ^h be the Hausdorff measure corresponding to a Hausdorff function $h \in \mathcal{H}$. Then all Borel subsets of X are μ^h-measurable.*

For any $n \in \mathbb{N}$, the function $h : [0, \infty) \to [0, \infty)$, defined by $h(t) = t^n$ for all $t \geq 0$, is a Hausdorff function.

Definition 3.42. Let (X, d) be a metric space and let $h : [0, \infty) \to [0, \infty)$ be defined by $h(t) = t^n$ for all $t \geq 0$, where n is a positive integer. The Hausdorff measure corresponding to the Hausdorff function h is denoted by $\mu^{(n)}$ and it is called the *n-dimensional Hausdorff measure* or the *(n)-measure*.

The one-dimensional Hausdorff measure of a simple curve in $\mathbb{R}^n$ is equal to the length of the curve, and the two-dimensional Hausdorff measure of a Lebesgue-measurable subset of $\mathbb{R}^2$ is proportional to the area of the set. Thus, the concept of the Hausdorff measure generalizes the Lebesgue measure and its notions of length, area, and volume. Indeed, there are d-dimensional Hausdorff measures for every $d \geq 0$, where d need not necessarily be an integer. These measures are fundamental in the geometric measure theory.

Now we will establish the relationship between (n)-measure $\mu^{(n)}$ and Lebesgue measure μ^* in the Euclidean space $\mathbb{R}^n$.

Theorem 3.43. *For any $n \in \mathbb{N}$, there exists a real constant κ_n with $0 < \kappa_n < \infty$ such that*

$$\mu^{(n)}(E) = \kappa_n \mu^*(E)$$

for all subsets E of the Euclidean space $\mathbb{R}^n$.

Proof. (a) We consider the cube C_0 of all points $x = (x_1, x_2, \ldots, x_n)$ of $\mathbb{R}^n$ with

$$0 \leq x_i < 1 \quad (\text{for } i \in \{1, 2, \ldots, n\}).$$

For any $\delta > 0$, we select an $N \in \mathbb{N}$ with $N > \frac{1}{\delta}\sqrt{n}$ such that the cube C_0 is covered by the collection of N^n smaller cubes of all points $x = (x_1, x_2, \ldots, x_n)$ of $\mathbb{R}^n$ with

$$\frac{r_i - 1}{N} \leq x_i < \frac{r_i}{N} \quad (\text{for } i \in \{1, 2, \ldots, n\}),$$

where $r_1, r_2, \ldots, r_n \in \{1, 2, \ldots, N\}$.

Since each of these smaller cubes has the diameter $\frac{1}{N}\sqrt{n}$ which is less than δ, it holds that

$$\mu_\delta^{(n)}(C_0) \leq \sum_{r_1=1}^{N} \sum_{r_2=1}^{N} \cdots \sum_{r_n=1}^{N} \left(\frac{1}{N}\sqrt{n}\right)^n = \sqrt{n}^n.$$

Thus, we have

$$\mu^{(n)}(C_0) \leq \sqrt{n}^n.$$

(b) We assume that $\{S_i : i \in \mathbb{N}\}$ is an arbitrary covering of C_0. For every $i \in \mathbb{N}$, we select a closed cube C_i containing S_i with the edge of C_i equal to twice the diameter of S_i. Then, we have

$$C_0 \subset \bigcup_{i=1}^{\infty} S_i \subset \bigcup_{i=1}^{\infty} C_i$$

and

$$1 = \mu^*(C_0) \leq \mu^*\left(\bigcup_{i=1}^{\infty} C_i\right) \leq \sum_{i=1}^{\infty} \mu^*(C_i) = \sum_{i=1}^{\infty} (2d(S_i))^n.$$

Hence, we get

$$\sum_{i=1}^{\infty} d(S_i)^n \geq \frac{1}{2^n}.$$

Since the inequality above is true for all coverings $\{S_i : i \in \mathbb{N}\}$ of C_0, we conclude that

$$\mu^{(n)}(C_0) \geq \frac{1}{2^n}.$$

(c) We now define

$$\kappa_n = \mu^{(n)}(C_0)$$

and we have

$$\frac{1}{2^n} \leq \kappa_n \leq \sqrt{n}^n.$$

Then, since $\mu^*(C_0) = 1$, we get

$$\mu^{(n)}(C_0) = \kappa_n \mu^*(C_0).$$

Since $\mu^{(n)}$ and μ^* are invariant under translation and they are homogeneous of degree n under similarity transformations, it holds that

$$\mu^{(n)}(C) = \kappa_n \mu^*(C) \tag{3.20}$$

for any cube C of all points $x = (x_1, x_2, \ldots, x_n)$ of $\mathbb{R}^n$ with

$$a_i \leq x_i < a_i + s \quad (\text{for } i \in \{1, 2, \ldots, n\}),$$

where $a_i \in \mathbb{R}$ for all $i \in \{1, 2, \ldots, n\}$ and s is an arbitrary real number with $0 < s < \infty$.

(d) We now choose a special collection $\mathcal{C}$ of all cubes of all points $x = (x_1, x_2, \ldots, x_n)$ of $\mathbb{R}^n$ with

$$\frac{r_i - 1}{2^k} \leq x_i < \frac{r_i}{2^k} \quad (\text{for } k \in \mathbb{N}_0,\ r_i \in \mathbb{Z},\ i \in \{1, 2, \ldots, n\}).$$

Each cube in $\mathcal{C}$ has the property that if two cubes have a point in common, one is contained in the other.

(e) Each point g of an arbitrary open subset G of $\mathbb{R}^n$ lies in a cube C of $\mathcal{C}$ that is contained in G and so in a cube C of $\mathcal{C}$ which is maximal in the sense that it is a cube C of $\mathcal{C}$ contained in G but not contained in any larger cube C' of $\mathcal{C}$ contained in G. Therefore, G is the union of the maximal cubes of $\mathcal{C}$ contained in G. According to the property mentioned in (d), these maximal cubes are obviously disjoint. Hence, we have

$$G = \bigcup_{i=1}^{\infty} C_i,$$

where the C_i's are disjoint cubes in $\mathcal{C}$.

Since all Borel subsets of $\mathbb{R}^n$ are measurable for both $\mu^{(n)}$ and μ^* by Theorems 3.26 and 3.30, it follows from (3.20) that

$$\mu^{(n)}(G) = \sum_{i=1}^{\infty} \mu^{(n)}(C_i) = \kappa_n \sum_{i=1}^{\infty} \mu^*(C_i) = \kappa_n \mu^*(G) \tag{3.21}$$

for any open subset G of $\mathbb{R}^n$.

(f) Assume that H is an arbitrary $\mathcal{G}_\delta$-set in $\mathbb{R}^n$. If $\mu^{(n)}(H) = \infty = \mu^*(H)$, then

$$\mu^{(n)}(H) = \kappa_n \mu^*(H).$$

If $\mu^{(n)}(H) < \infty$, then we can cover H by a sequence $\{G_i\}_{i \in \mathbb{N}}$ of open subsets of $\mathbb{R}^n$ with

$$\sum_{i=1}^{\infty} d(G_i)^n < \infty.$$

As we did in (b) (but using open cubes instead of closed cubes), we replace the open sets G_i with open cubes C_i of edge twice the diameter of G_i and construct an open covering $\{C_i : i \in \mathbb{N}\}$ of H such that

$$\mu^*\left(\bigcup_{i=1}^{\infty} C_i\right) < \infty.$$

On the other hand, when $\mu^*(H) < \infty$, H can be covered by an open subset of $\mathbb{R}^n$ with finite μ^*-measure. It follows from (3.21) that open sets of finite

μ^*-measure have finite $\mu^{(n)}$-measure. Hence, when H is a $\mathcal{G}_\delta$-set with either $\mu^{(n)}(H) < \infty$ or $\mu^*(H) < \infty$, H can be expressed as

$$H = \bigcap_{i=1}^{\infty} G_i,$$

where $\{G_i\}_{i\in\mathbb{N}}$ is a non-increasing sequence of measurable open sets of finite measure for both $\mu^{(n)}$ and μ^*. Thus, by (3.21) and Theorem 3.17 (ii), we have

$$\mu^{(n)}(H) = \inf\left\{\mu^{(n)}(G_i) : i \in \mathbb{N}\right\} = \kappa_n \inf\left\{\mu^*(G_i) : i \in \mathbb{N}\right\} = \kappa_n \mu^*(H)$$

for all $\mathcal{G}_\delta$-sets H.

(g) According to Corollary 3.32, $\mu^{(n)}$ and μ^* are $\mathcal{G}_\delta$-regular outer measures. If E is an arbitrary subset of $\mathbb{R}^n$, we can choose $\mathcal{G}_\delta$-sets H_1 and H_2 such that

$$E \subset H_1, \quad \mu^{(n)}(E) = \mu^{(n)}(H_1), \quad E \subset H_2, \quad \mu^*(E) = \mu^*(H_2).$$

Therefore, we conclude that

$$\mu^{(n)}(E) = \mu^{(n)}(H_1 \cap H_2) = \kappa_n \mu^*(H_1 \cap H_2) = \kappa_n \mu^*(E),$$

which completes the proof. □

Perhaps the most apt description of the relationship between $\mu^{(n)}$ and μ^* is that the former assigns unit measure to the ball of unit diameter, while the latter assigns unit measure to the n-cube of unit edge.

It is well known that $\mu^{(n)}(E) = \kappa_n \mu^*(E)$ for any subset E of the Euclidean space $\mathbb{R}^n$, where

$$\kappa_n = \left(\frac{2}{\sqrt{\pi}}\right)^n \Gamma\left(1 + \frac{n}{2}\right)$$

for all $n \in \mathbb{N}$.

Chapter 4

Extension of Isometries

In this chapter, we define the first- and second-order generalized spans and the index set, examine their properties, and apply them to the study of the extension of isometries. To this end, we develop a theory that extends the domain of local isometries to the generalized spans, where we call an isometry defined in a subset of a Hilbert space a local isometry. In addition, we prove that the domain of a local isometry can be extended to any real Hilbert space, where the domain of a local isometry does not have to be a convex body or an open set. To this purpose, among many references, S.-M. Jung's paper [9] is mainly cited.

4.1 Basic Concepts and Remarks

Throughout this chapter, using the symbol $\mathbb{R}^\omega$, we represent an infinite-dimensional real vector space defined as

$$\mathbb{R}^\omega = \big\{(x_1, x_2, \ldots) : x_i \in \mathbb{R} \text{ for all } i \in \mathbb{N}\big\}.$$

From now on, we denote by $(\mathbb{R}^\omega, \mathcal{T})$ the product space $\prod_{i=1}^{\infty} \mathbb{R}$, where $(\mathbb{R}, \mathcal{T}_\mathbb{R})$ is the usual topological space. Since $(\mathbb{R}, \mathcal{T}_\mathbb{R})$ is a Hausdorff space, it follows from Theorem 1.35 that the product space $(\mathbb{R}^\omega, \mathcal{T})$ is also a Hausdorff space.

Let $I^\omega = \prod_{i=1}^{\infty} I$ be the *Hilbert cube*, where $I = [0, 1]$ is the unit closed interval. We denote by $(I^\omega, \mathcal{T}_\omega)$ the (topological) subspace of $(\mathbb{R}^\omega, \mathcal{T})$. Then, $\mathcal{T}_\omega$ is the relative topology for I^ω induced by $\mathcal{T}$.

S.-M. Jung, *Ulam's Conjecture on Invariance of Measure in the Hilbert Cube*, Frontiers in Mathematics, https://doi.org/10.1007/978-3-031-30886-4_4

In this book, let $a = \{a_i\}_{i\in\mathbb{N}}$ be a sequence of positive real numbers satisfying the condition

$$\sum_{i=1}^{\infty} a_i^2 < \infty. \tag{4.1}$$

Using this sequence $a = \{a_i\}_{i\in\mathbb{N}}$, we define the metric on I^ω by

$$d_a(x,y) = \left(\sum_{i=1}^{\infty} a_i^2 (x_i - y_i)^2 \right)^{1/2} \tag{4.2}$$

for all $x = (x_1, x_2, \ldots) \in I^\omega$ and $y = (y_1, y_2, \ldots) \in I^\omega$.

Remark 4.1. It is to be noted that

(i) d_a is consistent with the topology $\mathcal{T}_\omega$ and it is translation-invariant (see [17]).

(ii) $(I^\omega, \mathcal{T}_\omega)$ is a Hausdorff space as a subspace of the Hausdorff space $(\mathbb{R}^\omega, \mathcal{T})$.

(iii) $(I^\omega, \mathcal{T}_\omega)$ is a compact subspace of $(\mathbb{R}^\omega, \mathcal{T})$ by Tychonoff's theorem.

We define

$$M_a = \left\{ (x_1, x_2, \ldots) \in \mathbb{R}^\omega : \sum_{i=1}^{\infty} a_i^2 x_i^2 < \infty \right\},$$

where $a = \{a_i\}_{i\in\mathbb{N}}$ is a sequence of positive real numbers that satisfies the condition (4.1). Then M_a is a vector space over $\mathbb{R}$, and we can define an inner product $\langle \cdot, \cdot \rangle_a$ on M_a by

$$\langle x, y \rangle_a = \sum_{i=1}^{\infty} a_i^2 x_i y_i$$

for all $x = (x_1, x_2, \ldots)$ and $y = (y_1, y_2, \ldots)$ of M_a. This inner product induces the norm

$$\|x\|_a = \sqrt{\langle x, x \rangle_a}$$

for all $x \in M_a$.

Remark 4.2. M_a is the set of all elements $x \in \mathbb{R}^\omega$ satisfying $\|x\|_a^2 < \infty$, i.e.,

$$M_a = \left\{ (x_1, x_2, \ldots) \in \mathbb{R}^\omega : \|x\|_a^2 < \infty \right\}.$$

In view of definition (4.2), the metric d_a on I^ω can be extended to the metric on M_a, i.e.,

$$d_a(x, y) = \sqrt{\langle x - y, x - y \rangle_a}$$

for all $x, y \in M_a$.

Similarly as [10, Theorem 2.1] and [13, Theorem 70.4], we prove the following theorem.

Theorem 4.3. *Assume that the sequence $a = \{a_i\}_{i\in\mathbb{N}}$ satisfies the condition* (4.1). *The inner product space $(M_a, \langle\cdot,\cdot\rangle_a)$ is complete, i.e., it is a real Hilbert space.*

Proof. Assume that $\{x_i\}_{i\in\mathbb{N}}$ is an arbitrary Cauchy sequence in M_a, where we set

$$x_i = (x_{i,1}, x_{i,2}, x_{i,3}, \ldots) \in M_a$$

for all $i \in \mathbb{N}$.

(a) Since $\{x_i\}_{i\in\mathbb{N}}$ is a Cauchy sequence in M_a, for any $\varepsilon > 0$, there exists an $N_\varepsilon \in \mathbb{N}$ such that $d_a(x_m, x_n) < \varepsilon$ for all integers $m, n \geq N_\varepsilon$. If we fix a $j \in \mathbb{N}$, it follows from the above fact that

$$\left|a_j x_{m,j} - a_j x_{n,j}\right| \leq \left(\sum_{k=1}^{\infty} a_k^2 (x_{m,k} - x_{n,k})^2\right)^{1/2} = d_a(x_m, x_n) < \varepsilon$$

for all integers $m, n \geq N_\varepsilon$, where a_j is a fixed positive real number as the jth term of the sequence $a = \{a_i\}_{i\in\mathbb{N}}$. This consideration implies that $\{x_{i,j}\}_{i\in\mathbb{N}}$ is a Cauchy sequence in $\mathbb{R}$ for each fixed $j \in \mathbb{N}$. Since $\mathbb{R}$ is complete, there exists a real number y_j such that

$$y_j = \lim_{i\to\infty} x_{i,j} \tag{4.3}$$

for every $j \in \mathbb{N}$. We now set

$$y = (y_1, y_2, y_3, \ldots) \in \mathbb{R}^\omega.$$

(b) Since $\{x_i\}_{i\in\mathbb{N}}$ is a Cauchy sequence in M_a, for any $\varepsilon > 0$, there exists an $N_\varepsilon \in \mathbb{N}$ such that $d_a(x_n, x_{n+p}) < \varepsilon$ for any $n, p \in \mathbb{N}$ with $n \geq N_\varepsilon$. Then, this fact implies that

$$\sum_{k=1}^{\infty} a_k^2 (x_{n,k} - x_{n+p,k})^2 = d_a(x_n, x_{n+p})^2 < \varepsilon^2$$

and hence, we have

$$\sum_{k=1}^{m} a_k^2 (x_{n,k} - x_{n+p,k})^2 < \varepsilon^2 \tag{4.4}$$

for all $m, n, p \in \mathbb{N}$ with $n \geq N_\varepsilon$. By (4.3) and (4.4), we get

$$\sum_{k=1}^{m} a_k^2 (x_{n,k} - y_k)^2 = \lim_{p \to \infty} \sum_{k=1}^{m} a_k^2 (x_{n,k} - x_{n+p,k})^2 \leq \varepsilon^2$$

for any $m, n \in \mathbb{N}$ with $n \geq N_\varepsilon$. Hence, we see that

$$\|x_n - y\|_a^2 = d_a(x_n, y)^2 = \sum_{k=1}^{\infty} a_k^2 (x_{n,k} - y_k)^2 \leq \varepsilon^2 \tag{4.5}$$

for all integers $n \geq N_\varepsilon$, i.e., we see by Remark 4.2 that $x_n - y \in M_a$ for each integer $n \geq N_\varepsilon$ and consequently

$$y \in M_a. \tag{4.6}$$

(c) Finally, it follows from (4.5) that, for each $\varepsilon > 0$, there exists an $N_\varepsilon \in \mathbb{N}$ such that $\|x_n - y\|_a \leq \varepsilon$ for any integer $n \geq N_\varepsilon$. In view of (4.6), this fact implies that each Cauchy sequence in M_a converges in M_a, i.e., the inner product space M_a is complete. □

What follows is a basic definition we are familiar with, but for the sake of completeness of the book, we now define precisely the d_a-isometry between subsets of M_a.

Definition 4.4. Let E_1 and E_2 be nonempty subsets of M_a.

(i) A function $f : E_1 \to E_2$ is called a d_a-*isometry* provided $d_a(f(x), f(y)) = d_a(x, y)$ for all $x, y \in E_1$.

(ii) E_1 is said to be d_a-*isometric* to E_2 provided there exists a surjective d_a-isometry $f : E_1 \to E_2$.

Let $(M_a, \mathcal{T}_a)$ be the topological space generated by the metric d_a. In view of Remark 4.1 $(ii), (iii)$ and using Theorem 1.43, it is easy to prove the following remarks. In view of Theorem 1.35, the topological space $(\mathbb{R}^\omega, \mathcal{T})$ is a Hausdorff space. Since $[-c, c]^\omega \subset M_a$ for any fixed $c > 0$, we may consider the families of open sets which are included in M_a only to prove Remark 4.5 (iii). This idea, together with Remark 4.1 (iii), implies the validity of Remark 4.5 (iii).

Remark 4.5. We note that

(i) $(M_a, \langle \cdot, \cdot \rangle_a)$ is a Hilbert space over $\mathbb{R}$.

(ii) $(M_a, \mathcal{T}_a)$ is a Hausdorff space as a subspace of the Hausdorff space $(\mathbb{R}^\omega, \mathcal{T})$.

(*iii*) $(I^\omega, \mathcal{T}_\omega)$ is a compact subspace of $(M_a, \mathcal{T}_a)$.

(*iv*) $(I^\omega, \mathcal{T}_\omega)$ is a closed subset of $(M_a, \mathcal{T}_a)$.

Definition 4.6. Given $c \in M_a$ the *translation* by c is the mapping $T_c : M_a \to M_a$ defined by $T_c(x) = x + c$ for all $x \in M_a$.

4.2 First-Order Generalized Span

In [10, Theorem 2.5], we were able to extend the domain of a d_a-isometry f to the whole space when the domain of f is a *non-degenerate basic cylinder* (see Definition 4.20 for the exact definition of non-degenerate basic cylinders). However, we shall see in Definition 4.33 and Theorem 4.34 that the domain of a d_a-isometry f can be extended to the whole space whenever f is defined on a bounded set which contains more than one element.

From now on, it is assumed that E, E_1, and E_2 are subsets of M_a, each of them contains more than one element, unless specifically stated for their cardinalities, and that they are bounded because the Hilbert cube I^ω is a bounded subset of M_a and the involved sets are indeed (Borel) subsets of I^ω in the main theorems.

If the set has only one element or no element, this case will not be covered here because the results derived from this case are trivial and uninteresting.

Definition 4.7. Assume that E is a nonempty bounded subset of M_a and p is a fixed element of E. We define the *first-order generalized span* of E with respect to p as

$$\mathrm{GS}(E,p) = \left\{ p + \sum_{i=1}^{m} \sum_{j=1}^{\infty} \alpha_{ij}(x_{ij} - p) \in M_a : m \in \mathbb{N}; \right.$$

$$\left. x_{ij} \in E \text{ and } \alpha_{ij} \in \mathbb{R} \text{ for all } i \text{ and } j \right\}.$$

We remark that if a bounded subset E of M_a contains more than one element, then E is a proper subset of its first-order generalized span $\mathrm{GS}(E,p)$, because $x = p + (x - p) \in \mathrm{GS}(E,p)$ for any $x \in E$ and $p + \alpha(x - p) \in \mathrm{GS}(E,p)$ for any $\alpha \in \mathbb{R}$, which implies that $\mathrm{GS}(E,p)$ is unbounded. Moreover, we note that $\alpha x + \beta y \in M_a$ for all $x, y \in M_a$ and $\alpha, \beta \in \mathbb{R}$, because $\|\alpha x + \beta y\|_a \leq |\alpha|\|x\|_a + |\beta|\|y\|_a < \infty$. Therefore, $\mathrm{GS}(E,p) - p$ is a real vector space, because the double sum in the definition of $\mathrm{GS}(E,p)$ guarantees $\alpha x + \beta y \in \mathrm{GS}(E,p) - p$ for all $x, y \in \mathrm{GS}(E,p) - p$ and $\alpha, \beta \in \mathbb{R}$ and because $\mathrm{GS}(E,p) - p$ is a subset of a real vector space M_a (*cf.* Lemma 4.17 (*i*) below).

We remark that the smallest flat containing E was introduced in [4] for any subset E of an n-dimensional Euclidean space somewhat similarly to the first-order generalized span as follows:

$$\begin{aligned} H(E) &= \left\{ \sum_{i=1}^{n} \alpha_i x_i : n \in \mathbb{N};\ x_i \in E \text{ and } \alpha_i \in \mathbb{R} \text{ for all } i \text{ with } \sum_{i=1}^{n} \alpha_i = 1 \right\} \\ &= \left\{ p + \sum_{i=1}^{n} \alpha_i (x_i - p) : n \in \mathbb{N};\ p, x_i \in E \text{ and } \alpha_i \in \mathbb{R} \text{ for all } i \right\}. \end{aligned}$$

Given a $p = (p_1, p_2, \ldots) \in M_a$ and an $n \in \mathbb{N}$, if we set

$$\mathbb{R}_p^n = \big\{ (x_1, x_2, \ldots) \in M_a : x_i \in \mathbb{R} \text{ for } 1 \leq i \leq n \text{ and } x_i = p_i \text{ for } i > n \big\},$$

then it is easy to see that $H(E) \subset \mathrm{GS}(E, p)$ for any set $E \subset \mathbb{R}_p^n$. However, it is obvious that $\mathrm{GS}(E, p) \not\subset H(E)$ for some $E \subset \mathbb{R}_p^n$.

For each $i \in \mathbb{N}$, we set $e_i = (0, \ldots, 0, 1, 0, \ldots)$, where 1 is in the ith position. Then $\big\{ \frac{1}{a_i} e_i \big\}_{i \in \mathbb{N}}$ is a complete orthonormal sequence in M_a. The following definition introduces the concepts of index and β-index based on the "standard" coordinate system $\big\{ \frac{1}{a_i} e_i \big\}_{i \in \mathbb{N}}$ and another one $\{\beta_i\}_{i \in \mathbb{N}}$, respectively.

Definition 4.8. Let E be a nonempty subset of M_a.

(i) We define the *index set* of E by

$$\Lambda(E) = \big\{ i \in \mathbb{N} : \text{there are an } x \in E \text{ and an } \alpha \in \mathbb{R} \setminus \{0\} \text{ satisfying } x + \alpha e_i \in E \big\}.$$

Each $i \in \Lambda(E)$ is called an *index* of E. If $\Lambda(E) \neq \mathbb{N}$, then the set E is called *degenerate*. Otherwise, E is called *non-degenerate*.

(ii) Let $\beta = \{\beta_i\}_{i \in \mathbb{N}}$ be another complete orthonormal sequence in M_a. We define the *β-index set* of E by

$$\Lambda_\beta(E) = \big\{ i \in \mathbb{N} : \text{there are an } x \in E \text{ and an } \alpha \in \mathbb{R} \setminus \{0\} \text{ satisfying } x + \alpha \beta_i \in E \big\}.$$

Each $i \in \Lambda_\beta(E)$ is called a *β-index* of E.

Here we should not be confused with the terminologies "index" and "index set" used in Chap. 1. Because the meanings used in Chap. 1 are very different, there will be no confusion even if we use the terms in Definition 4.8. Therefore, we will continue to use the terminologies given in the above definition.

We will find that the concept of index set in Hilbert space sometimes takes over the role that the concept of dimension plays in vector space. According to the definition above, if i is an index of E, i.e., $i \in \Lambda(E)$, then there are $x \in E$ and $x + \alpha e_i \in E$ for some $\alpha \neq 0$. Since $x \neq x + \alpha e_i$, we remark that if $\Lambda(E) \neq \emptyset$, then the set E contains at least two elements.

In the following lemma, we prove that if i is an index of E and $p \in E$, then the first-order generalized span $\mathrm{GS}(E, p)$ contains the line through p in the direction e_i.

Lemma 4.9. *Let $\beta = \{\beta_i\}_{i \in \mathbb{N}}$ be a complete orthonormal sequence in M_a. Assume that E is a bounded subset of M_a and $\mathrm{GS}(E, p)$ is the first-order generalized span of E with respect to a fixed element $p \in E$. If $i \in \Lambda_\beta(E)$, then $p + \alpha\beta_i \in \mathrm{GS}(E, p)$ for all $\alpha \in \mathbb{R}$.*

Proof. By Definition 4.8 (ii), if $i \in \Lambda_\beta(E)$ then there exists an $x \in E$ and an $\alpha_0 \neq 0$, which satisfy $x + \alpha_0\beta_i \in E$. Since $x \in E$ and $x + \alpha_0\beta_i \in E$, by Definition 4.7, we get

$$p + \alpha_0\gamma\beta_i = p + \gamma(x + \alpha_0\beta_i - p) - \gamma(x - p) \in \mathrm{GS}(E, p)$$

for all $\gamma \in \mathbb{R}$. Setting $\alpha = \alpha_0\gamma$ in the above relation, we obtain $p + \alpha\beta_i \in \mathrm{GS}(E, p)$ for any $\alpha \in \mathbb{R}$. □

We now introduce a lemma, which is a generalized version of [10, Lemma 2.3] and whose proof runs in the same way. We prove that the function $T_{-q} \circ f \circ T_p : E_1 - p \to E_2 - q$ preserves the inner product. This property is important for proving several theorems in this chapter as a necessary condition for f to be a d_a-isometry.

Lemma 4.10. *Assume that E_1 and E_2 are bounded subsets of M_a that are d_a-isometric to each other via a surjective d_a-isometry $f : E_1 \to E_2$. Assume that p is an element of E_1 and q is an element of E_2 with $q = f(p)$. Then the function $T_{-q} \circ f \circ T_p : E_1 - p \to E_2 - q$ preserves the inner product, i.e.,*

$$\big\langle (T_{-q} \circ f \circ T_p)(x - p), (T_{-q} \circ f \circ T_p)(y - p) \big\rangle_a = \langle x - p, y - p \rangle_a$$

for all $x, y \in E_1$.

Proof. Since $T_{-q} \circ f \circ T_p : E_1 - p \to E_2 - q$ is a d_a-isometry, we have

$$\|(T_{-q} \circ f \circ T_p)(x - p) - (T_{-q} \circ f \circ T_p)(y - p)\|_a^2 = \|(x - p) - (y - p)\|_a^2$$

for any $x, y \in E_1$. If we put $y = p$ in the last equality, then we get

$$\|(T_{-q} \circ f \circ T_p)(x - p)\|_a^2 = \|x - p\|_a^2$$

for each $x \in E_1$. Moreover, it follows from the previous equality that

$$\begin{aligned}
&\|(T_{-q} \circ f \circ T_p)(x-p) - (T_{-q} \circ f \circ T_p)(y-p)\|_a^2 \\
&\quad = \big\langle (T_{-q} \circ f \circ T_p)(x-p) - (T_{-q} \circ f \circ T_p)(y-p), \\
&\qquad\quad (T_{-q} \circ f \circ T_p)(x-p) - (T_{-q} \circ f \circ T_p)(y-p) \big\rangle_a \\
&\quad = \|x-p\|_a^2 - 2\big\langle (T_{-q} \circ f \circ T_p)(x-p), (T_{-q} \circ f \circ T_p)(y-p) \big\rangle_a + \|y-p\|_a^2
\end{aligned}$$

and

$$\begin{aligned}
\|(x-p)-(y-p)\|_a^2 &= \big\langle (x-p)-(y-p), (x-p)-(y-p) \big\rangle_a \\
&= \|x-p\|_a^2 - 2\langle x-p, y-p\rangle_a + \|y-p\|_a^2.
\end{aligned}$$

Finally, comparing the last two equalities yields the validity of our assertion. □

4.3 First-Order Extension of Isometries

In the last section, we made all the necessary preparations to extend the domain E_1 of the surjective d_a-isometry $f : E_1 \to E_2$ to its first-order generalized span $\mathrm{GS}(E_1, p)$.

Although E_1 is a bounded subset of M_a, the translation of its first-order generalized span by $-p$, $\mathrm{GS}(E_1, p) - p$, is a real vector space. Now we will extend the d_a-isometry $T_{-q} \circ f \circ T_p$ defined on the bounded set $E_1 - p$ to the d_a-isometry $T_{-q} \circ F \circ T_p$ defined on the vector space $\mathrm{GS}(E_1, p) - p$. Comparing their "sizes" of $E_1 - p$ and $\mathrm{GS}(E_1, p) - p$, or considering that $\mathrm{GS}(E_1, p) - p$ is an algebraically closed space, it is a great achievement to extend the d_a-isometry $T_{-q} \circ f \circ T_p$ defined on the bounded set $E_1 - p$ to the d_a-isometry defined on the vector space $\mathrm{GS}(E_1, p) - p$.

Definition 4.11. Assume that E_1 and E_2 are nonempty bounded subsets of M_a that are d_a-isometric to each other via a surjective d_a-isometry $f : E_1 \to E_2$. Let p be a fixed element of E_1 and let q be an element of E_2 that satisfies $q = f(p)$. We define a function $F : \mathrm{GS}(E_1, p) \to M_a$ as

$$\begin{aligned}
&(T_{-q} \circ F \circ T_p)\left(\sum_{i=1}^{m} \sum_{j=1}^{\infty} \alpha_{ij}(x_{ij} - p) \right) \\
&\quad = \sum_{i=1}^{m} \sum_{j=1}^{\infty} \alpha_{ij}(T_{-q} \circ f \circ T_p)(x_{ij} - p)
\end{aligned}$$

for any $m \in \mathbb{N}$, $x_{ij} \in E_1$, and for all $\alpha_{ij} \in \mathbb{R}$ satisfying $\sum_{i=1}^{m} \sum_{j=1}^{\infty} \alpha_{ij}(x_{ij} - p) \in M_a$.

We note that in the definition above, it is important for the argument of $T_{-q} \circ F \circ T_p$ to belong to M_a. Now we show that the function $F : \mathrm{GS}(E_1, p) \to M_a$ is well defined.

Lemma 4.12. *Assume that E_1 and E_2 are bounded subsets of M_a that are d_a-isometric to each other via a surjective d_a-isometry $f : E_1 \to E_2$. Let p be an element of E_1 and let q be an element of E_2 that satisfy $q = f(p)$. The function $F : \mathrm{GS}(E_1, p) \to M_a$ given in Definition 4.11 is well defined.*

Proof. First, we will check that the range of F is a subset of M_a. For any $m, n_1, n_2 \in \mathbb{N}$ with $n_2 > n_1$, $x_{ij} \in E_1$, and for all $\alpha_{ij} \in \mathbb{R}$, it follows from Lemma 4.10 that

$$\begin{aligned}
&\left\| \sum_{i=1}^{m} \sum_{j=1}^{n_2} \alpha_{ij} (T_{-q} \circ f \circ T_p)(x_{ij} - p) - \sum_{i=1}^{m} \sum_{j=1}^{n_1} \alpha_{ij} (T_{-q} \circ f \circ T_p)(x_{ij} - p) \right\|_a^2 \\
&= \left\langle \sum_{i=1}^{m} \sum_{j=n_1+1}^{n_2} \alpha_{ij} (T_{-q} \circ f \circ T_p)(x_{ij} - p), \right. \\
&\qquad \left. \sum_{k=1}^{m} \sum_{\ell=n_1+1}^{n_2} \alpha_{k\ell} (T_{-q} \circ f \circ T_p)(x_{k\ell} - p) \right\rangle_a \\
&= \sum_{i=1}^{m} \sum_{k=1}^{m} \sum_{j=n_1+1}^{n_2} \alpha_{ij} \\
&\qquad \sum_{\ell=n_1+1}^{n_2} \alpha_{k\ell} \Big\langle (T_{-q} \circ f \circ T_p)(x_{ij} - p), \, (T_{-q} \circ f \circ T_p)(x_{k\ell} - p) \Big\rangle_a \\
&= \sum_{i=1}^{m} \sum_{k=1}^{m} \sum_{j=n_1+1}^{n_2} \alpha_{ij} \sum_{\ell=n_1+1}^{n_2} \alpha_{k\ell} \langle x_{ij} - p, \, x_{k\ell} - p \rangle_a \\
&= \left\langle \sum_{i=1}^{m} \sum_{j=n_1+1}^{n_2} \alpha_{ij} (x_{ij} - p), \sum_{k=1}^{m} \sum_{\ell=n_1+1}^{n_2} \alpha_{k\ell} (x_{k\ell} - p) \right\rangle_a .
\end{aligned}$$

Hence, we have

$$\begin{aligned}
&\left\| \sum_{i=1}^{m} \sum_{j=1}^{n_2} \alpha_{ij} (T_{-q} \circ f \circ T_p)(x_{ij} - p) - \sum_{i=1}^{m} \sum_{j=1}^{n_1} \alpha_{ij} (T_{-q} \circ f \circ T_p)(x_{ij} - p) \right\|_a^2 \\
&= \left\| \sum_{i=1}^{m} \sum_{j=n_1+1}^{n_2} \alpha_{ij} (x_{ij} - p) \right\|_a^2 \qquad (4.7) \\
&= \left\| \sum_{i=1}^{m} \sum_{j=1}^{n_2} \alpha_{ij} (x_{ij} - p) - \sum_{i=1}^{m} \sum_{j=1}^{n_1} \alpha_{ij} (x_{ij} - p) \right\|_a^2 .
\end{aligned}$$

Indeed, the equality (4.7) holds for all $m, n_1, n_2 \in \mathbb{N}$.

We now assume that $\sum_{i=1}^{m} \sum_{j=1}^{\infty} \alpha_{ij}(x_{ij}-p) \in M_a$ for some $x_{ij} \in E_1$ and $\alpha_{ij} \in \mathbb{R}$, where m is a fixed positive integer. Since $(M_a, \mathcal{T}_a)$ is a Hausdorff space on account of Remark 4.5 (ii) and the topology $\mathcal{T}_a$ is consistent with the metric d_a and with the norm $\|\cdot\|_a$ (*cf.* Remark 4.1 (i)), the sequence $\left\{ \sum_{i=1}^{m} \sum_{j=1}^{n} \alpha_{ij}(x_{ij} - p) \right\}_n$ converges to $\sum_{i=1}^{m} \sum_{j=1}^{\infty} \alpha_{ij}(x_{ij} - p)$ (in M_a) and hence, the sequence $\left\{ \sum_{i=1}^{m} \sum_{j=1}^{n} \alpha_{ij}(x_{ij} - p) \right\}_n$ is a Cauchy sequence in M_a.

We know by (4.7) and the definition of Cauchy sequences that for each $\varepsilon > 0$ there exists an integer $N_\varepsilon > 0$ such that

$$\begin{aligned} &\left\| \sum_{i=1}^{m} \sum_{j=1}^{n_2} \alpha_{ij}(T_{-q} \circ f \circ T_p)(x_{ij} - p) - \sum_{i=1}^{m} \sum_{j=1}^{n_1} \alpha_{ij}(T_{-q} \circ f \circ T_p)(x_{ij} - p) \right\|_a \\ &\quad = \left\| \sum_{i=1}^{m} \sum_{j=1}^{n_2} \alpha_{ij}(x_{ij} - p) - \sum_{i=1}^{m} \sum_{j=1}^{n_1} \alpha_{ij}(x_{ij} - p) \right\|_a < \varepsilon \end{aligned}$$

for all $n_1, n_2 > N_\varepsilon$, which implies that $\left\{ \sum_{i=1}^{m} \sum_{j=1}^{n} \alpha_{ij}(T_{-q} \circ f \circ T_p)(x_{ij} - p) \right\}_n$ is also a Cauchy sequence in M_a. As we proved in [10, Theorem 2.1] or by Remark 4.5 (i), we observe that $(M_a, \langle \cdot, \cdot \rangle_a)$ is a real Hilbert space when the sequence $a = \{a_i\}_{i \in \mathbb{N}}$ satisfies the condition (4.1). Thus, M_a is complete, so the Cauchy sequence $\left\{ \sum_{i=1}^{m} \sum_{j=1}^{n} \alpha_{ij}(T_{-q} \circ f \circ T_p)(x_{ij} - p) \right\}_n$ converges in M_a, i.e., by Definition 4.11, we have

$$\begin{aligned} &(T_{-q} \circ F \circ T_p)\left(\sum_{i=1}^{m} \sum_{j=1}^{\infty} \alpha_{ij}(x_{ij} - p) \right) \\ &\quad = \sum_{i=1}^{m} \sum_{j=1}^{\infty} \alpha_{ij}(T_{-q} \circ f \circ T_p)(x_{ij} - p) \\ &\quad = \lim_{n \to \infty} \sum_{i=1}^{m} \sum_{j=1}^{n} \alpha_{ij}(T_{-q} \circ f \circ T_p)(x_{ij} - p) \\ &\quad \in M_a, \end{aligned}$$

which implies

$$F\left(p + \sum_{i=1}^{m} \sum_{j=1}^{\infty} \alpha_{ij}(x_{ij} - p) \right) \in M_a + q = M_a$$

for all $x_{ij} \in E_1$ and $\alpha_{ij} \in \mathbb{R}$ with $\sum_{i=1}^{m} \sum_{j=1}^{\infty} \alpha_{ij}(x_{ij} - p) \in M_a$, i.e., the image of each element of $\mathrm{GS}(E_1, p)$ under F belongs to M_a.

We now assume that $\sum_{i=1}^{m_1} \sum_{j=1}^{\infty} \alpha_{ij}(x_{ij} - p) = \sum_{i=1}^{m_2} \sum_{j=1}^{\infty} \beta_{ij}(y_{ij} - p) \in M_a$ for some $m_1, m_2 \in \mathbb{N}$, $x_{ij}, y_{ij} \in E_1$, and for some $\alpha_{ij}, \beta_{ij} \in \mathbb{R}$. It then follows from Definition 4.11 and Lemma 4.10 that

$$
\begin{aligned}
&\bigg\| (T_{-q} \circ F \circ T_p)\bigg(\sum_{i=1}^{m_1} \sum_{j=1}^{\infty} \alpha_{ij}(x_{ij} - p) \bigg) \\
&\qquad - (T_{-q} \circ F \circ T_p)\bigg(\sum_{i=1}^{m_2} \sum_{j=1}^{\infty} \beta_{ij}(y_{ij} - p) \bigg) \bigg\|_a^2 \\
&= \bigg\| \sum_{i=1}^{m_1} \sum_{j=1}^{\infty} \alpha_{ij}(T_{-q} \circ f \circ T_p)(x_{ij} - p) - \sum_{i=1}^{m_2} \sum_{j=1}^{\infty} \beta_{ij}(T_{-q} \circ f \circ T_p)(y_{ij} - p) \bigg\|_a^2 \\
&= \bigg\langle \sum_{i=1}^{m_1} \sum_{j=1}^{\infty} \alpha_{ij}(T_{-q} \circ f \circ T_p)(x_{ij} - p) - \sum_{i=1}^{m_2} \sum_{j=1}^{\infty} \beta_{ij}(T_{-q} \circ f \circ T_p)(y_{ij} - p), \\
&\qquad \sum_{k=1}^{m_1} \sum_{\ell=1}^{\infty} \alpha_{k\ell}(T_{-q} \circ f \circ T_p)(x_{k\ell} - p) - \sum_{k=1}^{m_2} \sum_{\ell=1}^{\infty} \beta_{k\ell}(T_{-q} \circ f \circ T_p)(y_{k\ell} - p) \bigg\rangle_a \\
&= \bigg\langle \sum_{i=1}^{m_1} \sum_{j=1}^{\infty} \alpha_{ij}(x_{ij} - p) - \sum_{i=1}^{m_2} \sum_{j=1}^{\infty} \beta_{ij}(y_{ij} - p), \\
&\qquad \sum_{k=1}^{m_1} \sum_{\ell=1}^{\infty} \alpha_{k\ell}(x_{k\ell} - p) - \sum_{k=1}^{m_2} \sum_{\ell=1}^{\infty} \beta_{k\ell}(y_{k\ell} - p) \bigg\rangle_a \\
&= \bigg\| \sum_{i=1}^{m_1} \sum_{j=1}^{\infty} \alpha_{ij}(x_{ij} - p) - \sum_{i=1}^{m_2} \sum_{j=1}^{\infty} \beta_{ij}(y_{ij} - p) \bigg\|_a^2 \\
&= 0,
\end{aligned}
$$

which implies that

$$
\begin{aligned}
&(T_{-q} \circ F \circ T_p)\bigg(\sum_{i=1}^{m_1} \sum_{j=1}^{\infty} \alpha_{ij}(x_{ij} - p) \bigg) \\
&\qquad = (T_{-q} \circ F \circ T_p)\bigg(\sum_{i=1}^{m_2} \sum_{j=1}^{\infty} \beta_{ij}(y_{ij} - p) \bigg)
\end{aligned}
$$

for all $m_1, m_2 \in \mathbb{N}$, $x_{ij}, y_{ij} \in E_1$, and for all $\alpha_{ij}, \beta_{ij} \in \mathbb{R}$ which satisfy the condition $\sum_{i=1}^{m_1} \sum_{j=1}^{\infty} \alpha_{ij}(x_{ij} - p) = \sum_{i=1}^{m_2} \sum_{j=1}^{\infty} \beta_{ij}(y_{ij} - p) \in M_a$. $\square$

In [6, Theorem 2.2], we were able to extend the domain of a d_a-isometry $f : J \to K$ to the whole space M_a when J is a non-degenerate basic cylinder, while we prove in the following theorem that the domain of a d_a-isometry $f : E_1 \to E_2$ can be extended to the first-order generalized span $\mathrm{GS}(E_1, p)$ whenever E_1 is a nonempty bounded subset of M_a, whether degenerate or non-degenerate. Therefore, Theorem 4.13 is a generalization of [6, Theorem 2.2].

In the proof, we use the fact that $\mathrm{GS}(E_1, p) - p$ is a real vector space. This fact is self-evident, as briefly mentioned earlier.

Theorem 4.13. *Assume that E_1 and E_2 are bounded subsets of M_a that are d_a-isometric to each other via a surjective d_a-isometry $f : E_1 \to E_2$. Assume that p is an element of E_1 and q is an element of E_2 with $q = f(p)$. The function $F : \mathrm{GS}(E_1, p) \to M_a$ defined in Definition 4.11 is a d_a-isometry and the function $T_{-q} \circ F \circ T_p : \mathrm{GS}(E_1, p) - p \to M_a$ is a linear d_a-isometry. In particular, F is an extension of f.*

Proof. (a) Let u and v be arbitrary elements of the first-order generalized span $\mathrm{GS}(E_1, p)$ of E_1 with respect to p. Then

$$
\begin{aligned}
u - p &= \sum_{i=1}^{m} \sum_{j=1}^{\infty} \alpha_{ij}(x_{ij} - p) \in M_a, \\
v - p &= \sum_{i=1}^{n} \sum_{j=1}^{\infty} \beta_{ij}(y_{ij} - p) \in M_a
\end{aligned}
\tag{4.8}
$$

for some $m, n \in \mathbb{N}$, some $x_{ij}, y_{ij} \in E_1$, and for some $\alpha_{ij}, \beta_{ij} \in \mathbb{R}$. Then, according to Definition 4.11, we have

$$
\begin{aligned}
(T_{-q} \circ F \circ T_p)(u - p) &= \sum_{i=1}^{m} \sum_{j=1}^{\infty} \alpha_{ij}(T_{-q} \circ f \circ T_p)(x_{ij} - p), \\
(T_{-q} \circ F \circ T_p)(v - p) &= \sum_{i=1}^{n} \sum_{j=1}^{\infty} \beta_{ij}(T_{-q} \circ f \circ T_p)(y_{ij} - p).
\end{aligned}
\tag{4.9}
$$

(*b*) By Lemma 4.10, (4.8), and (4.9), we get

$$\begin{aligned}
&\big\langle (T_{-q} \circ F \circ T_p)(u-p), (T_{-q} \circ F \circ T_p)(v-p) \big\rangle_a \\
&= \left\langle \sum_{i=1}^{m} \sum_{j=1}^{\infty} \alpha_{ij} (T_{-q} \circ f \circ T_p)(x_{ij} - p), \right. \\
&\qquad \left. \sum_{k=1}^{n} \sum_{\ell=1}^{\infty} \beta_{k\ell} (T_{-q} \circ f \circ T_p)(y_{k\ell} - p) \right\rangle_a \\
&= \sum_{i=1}^{m} \sum_{k=1}^{n} \sum_{j=1}^{\infty} \alpha_{ij} \sum_{\ell=1}^{\infty} \beta_{k\ell} \Big\langle (T_{-q} \circ f \circ T_p)(x_{ij} - p), \\
&\qquad\qquad (T_{-q} \circ f \circ T_p)(y_{k\ell} - p) \Big\rangle_a \qquad (4.10) \\
&= \sum_{i=1}^{m} \sum_{k=1}^{n} \sum_{j=1}^{\infty} \alpha_{ij} \sum_{\ell=1}^{\infty} \beta_{k\ell} \langle x_{ij} - p, y_{k\ell} - p \rangle_a \\
&= \left\langle \sum_{i=1}^{m} \sum_{j=1}^{\infty} \alpha_{ij} (x_{ij} - p), \sum_{k=1}^{n} \sum_{\ell=1}^{\infty} \beta_{k\ell} (y_{k\ell} - p) \right\rangle_a \\
&= \langle u - p, v - p \rangle_a
\end{aligned}$$

for all $u, v \in \mathrm{GS}(E_1, p)$. That is, $T_{-q} \circ F \circ T_p$ preserves the inner product. Indeed, equality (4.10) is an extended version of Lemma 4.10.

(*c*) By using equality (4.10), we further obtain

$$\begin{aligned}
&d_a\big(F(u), F(v)\big)^2 \\
&= \|F(u) - F(v)\|_a^2 \\
&= \big\| (T_{-q} \circ F \circ T_p)(u-p) - (T_{-q} \circ F \circ T_p)(v-p) \big\|_a^2 \\
&= \big\langle (T_{-q} \circ F \circ T_p)(u-p) - (T_{-q} \circ F \circ T_p)(v-p), \\
&\qquad (T_{-q} \circ F \circ T_p)(u-p) - (T_{-q} \circ F \circ T_p)(v-p) \big\rangle_a \\
&= \langle u-p, u-p \rangle_a - \langle u-p, v-p \rangle_a - \langle v-p, u-p \rangle_a + \langle v-p, v-p \rangle_a \\
&= \big\langle (u-p) - (v-p), (u-p) - (v-p) \big\rangle_a \\
&= \|(u-p) - (v-p)\|_a^2 \\
&= \|u - v\|_a^2 \\
&= d_a(u, v)^2
\end{aligned}$$

for all $u, v \in \mathrm{GS}(E_1, p)$, i.e., F is a d_a-isometry.

(*d*) Now, let u and v be arbitrary elements of $\mathrm{GS}(E_1, p)$. Then, it holds that $u - p \in \mathrm{GS}(E_1, p) - p$, $v - p \in \mathrm{GS}(E_1, p) - p$, and $\alpha(u-p) + \beta(v-p) \in \mathrm{GS}(E_1, p) - p$ for any $\alpha, \beta \in \mathbb{R}$, because $\mathrm{GS}(E_1, p) - p$ is a real vector space.

We get

$$\begin{aligned}
&\big\|(T_{-q}\circ F\circ T_p)\big(\alpha(u-p)+\beta(v-p)\big)\\
&\quad -\alpha(T_{-q}\circ F\circ T_p)(u-p)-\beta(T_{-q}\circ F\circ T_p)(v-p)\big\|_a^2\\
&\quad =\Big\langle (T_{-q}\circ F\circ T_p)\big(\alpha(u-p)+\beta(v-p)\big)\\
&\qquad\quad -\alpha(T_{-q}\circ F\circ T_p)(u-p)-\beta(T_{-q}\circ F\circ T_p)(v-p),\\
&\qquad\quad (T_{-q}\circ F\circ T_p)\big(\alpha(u-p)+\beta(v-p)\big)\\
&\qquad\quad -\alpha(T_{-q}\circ F\circ T_p)(u-p)-\beta(T_{-q}\circ F\circ T_p)(v-p)\Big\rangle_a.
\end{aligned}$$

Since $\alpha(u-p)+\beta(v-p)=w-p$ for some $w\in \mathrm{GS}(E_1,p)$, we further use (4.10) to obtain

$$\begin{aligned}
&\big\|(T_{-q}\circ F\circ T_p)\big(\alpha(u-p)+\beta(v-p)\big)\\
&\quad -\alpha(T_{-q}\circ F\circ T_p)(u-p)-\beta(T_{-q}\circ F\circ T_p)(v-p)\big\|_a^2\\
&\quad =\langle w-p,\,w-p\rangle_a-\alpha\langle w-p,\,u-p\rangle_a-\beta\langle w-p,\,v-p\rangle_a\\
&\qquad -\alpha\langle u-p,\,w-p\rangle_a+\alpha^2\langle u-p,\,u-p\rangle_a+\alpha\beta\langle u-p,\,v-p\rangle_a\\
&\qquad -\beta\langle v-p,\,w-p\rangle_a+\alpha\beta\langle v-p,\,u-p\rangle_a+\beta^2\langle v-p,\,v-p\rangle_a\\
&\quad =0,
\end{aligned}$$

which implies that the function $T_{-q}\circ F\circ T_p:\mathrm{GS}(E_1,p)-p\to M_a$ is linear.

(e) Finally, we set $\alpha_{11}=1$, $\alpha_{ij}=0$ for any $(i,j)\neq(1,1)$, and $x_{11}=x$ in (4.8) and (4.9) to see

$$(T_{-q}\circ F\circ T_p)(x-p)=(T_{-q}\circ f\circ T_p)(x-p)$$

for every $x\in E_1$, which implies that $F(x)=f(x)$ for every $x\in E_1$, i.e., F is an extension of f. □

4.4 Second-Order Generalized Span

For any element x of M_a and $r>0$, we denote by $B_r(x)$ the open ball defined by $B_r(x)=\{y\in M_a:\|y-x\|_a<r\}$.

Definitions 4.7 and 4.11 will be generalized to the cases of $n\geq 2$ in the following definition. We introduce the concept of nth-order generalized span $\mathrm{GS}^n(E_1,p)$, which generalizes the concept of first-order generalized span $\mathrm{GS}(E_1,p)$. Moreover, we define the d_a-isometry F_n which extends the domain E_1 of a d_a-isometry f to $\mathrm{GS}^n(E_1,p)$.

It is surprising, however, that this process of generalization does not go far. Indeed, we will find in Proposition 4.18 and Theorem 4.28 that $\mathrm{GS}^2(E_1, p)$ and F_2 are their limits.

Definition 4.14. Let E_1 be a nonempty bounded subset of M_a that is d_a-isometric to a subset E_2 of M_a via a surjective d_a-isometry $f : E_1 \to E_2$. Let p be an element of E_1 and q an element of E_2 with $q = f(p)$. Assume that r is a positive real number satisfying $E_1 \subset B_r(p)$.

(i) We define $\mathrm{GS}^0(E_1, p) = E_1$ and $\mathrm{GS}^1(E_1, p) = \mathrm{GS}(E_1, p)$. In general, we define the nth-*order generalized span* of E_1 with respect to p as $\mathrm{GS}^n(E_1, p) = \mathrm{GS}(\mathrm{GS}^{n-1}(E_1, p) \cap B_r(p), p)$ for all $n \in \mathbb{N}$.

(ii) We define $F_0 = f$ and $F_1 = F$, where F is defined in Definition 4.11. Moreover, for any $n \in \mathbb{N}$, we define the function $F_n : \mathrm{GS}^n(E_1, p) \to M_a$ by

$$(T_{-q} \circ F_n \circ T_p)\left(\sum_{i=1}^{m}\sum_{j=1}^{\infty} \alpha_{ij}(x_{ij} - p)\right) = \sum_{i=1}^{m}\sum_{j=1}^{\infty} \alpha_{ij}(T_{-q} \circ F_{n-1} \circ T_p)(x_{ij} - p)$$

for all $m \in \mathbb{N}$, $x_{ij} \in \mathrm{GS}^{n-1}(E_1, p) \cap B_r(p)$, and $\alpha_{ij} \in \mathbb{R}$ which satisfy the condition $\sum_{i=1}^{m}\sum_{j=1}^{\infty} \alpha_{ij}(x_{ij} - p) \in M_a$.

Let E be a nonempty bounded subset of M_a. It is not difficult to show that $\mathrm{GS}^n(E, p) - p$ is a real vector space for each $n \in \mathbb{N}$: Obviously, $\mathrm{GS}(E, p) - p$ is a real vector space. We assume that $\mathrm{GS}^n(E, p) - p$ is a real vector space for some $n \in \mathbb{N}$ and u, v are arbitrary elements of $\mathrm{GS}^{n+1}(E, p) - p$. Then there exist $m_1, m_2 \in \mathbb{N}$, $\alpha_{ij}, \beta_{ij} \in \mathbb{R}$ and $x_{ij}, y_{ij} \in \mathrm{GS}^n(E, p) \cap B_r(p)$, where r is some positive real constant with $E \subset B_r(p)$, such that

$$u = \sum_{i=1}^{m_1}\sum_{j=1}^{\infty} \alpha_{ij}(x_{ij} - p) \quad \text{and} \quad v = \sum_{i=1}^{m_2}\sum_{j=1}^{\infty} \beta_{ij}(y_{ij} - p).$$

Further, for any $\alpha, \beta \in \mathbb{R}$, we see that

$$\begin{aligned}\alpha u + \beta v &= \sum_{i=1}^{m_1}\sum_{j=1}^{\infty} \alpha\alpha_{ij}(x_{ij} - p) + \sum_{i=1}^{m_2}\sum_{j=1}^{\infty} \beta\beta_{ij}(y_{ij} - p) \\ &\in \mathrm{GS}^{n+1}(E, p) - p,\end{aligned}$$

which implies that $\mathrm{GS}^{n+1}(E,p)-p$ is a real vector space as a subspace of the real vector space M_a. Thus we see by induction conclusion that $\mathrm{GS}^n(E,p)-p$ is a real vector space for every $n \in \mathbb{N}$.

Proposition 4.15. *Assume that E is a nonempty bounded subset of M_a and $p \in E$. If s and t are positive real numbers that satisfy $E \subset B_s(p) \cap B_t(p)$, then*

$$\mathrm{GS}\big(\mathrm{GS}^n(E,p) \cap B_s(p), p\big) = \mathrm{GS}\big(\mathrm{GS}^n(E,p) \cap B_t(p), p\big)$$

for all $n \in \mathbb{N}$.

Proof. Assume that $0 < s < t$. Then, there exists a real number $c > 1$ with $s > \frac{t}{c}$ and it is obvious that $B_{t/c}(p) \subset B_s(p)$. Assume that x is an arbitrary element of $\mathrm{GS}(\mathrm{GS}^n(E,p) \cap B_t(p), p)$. Then there exist some $m \in \mathbb{N}$, some $u_{ij} \in \mathrm{GS}^n(E,p) \cap B_t(p)$ and some $\alpha_{ij} \in \mathbb{R}$ such that $x = p + \sum_{i=1}^{m} \sum_{j=1}^{\infty} \alpha_{ij}(u_{ij}-p) \in M_a$. We note that

$$\big(\mathrm{GS}^n(E,p)-p\big) \cap \big(B_t(p)-p\big) = \big\{u-p \in M_a : u \in \mathrm{GS}^n(E,p) \cap B_t(p)\big\}.$$

Since $\mathrm{GS}^n(E,p)-p$ is a real vector space, $\frac{t}{c} < s$, and since $u_{ij}-p \in (\mathrm{GS}^n(E,p)-p) \cap (B_t(p)-p)$ for any i and j, we have

$$\frac{1}{c}(u_{ij}-p) \in (\mathrm{GS}^n(E,p)-p) \cap (B_s(p)-p).$$

Hence, we can choose a $v_{ij} \in \mathrm{GS}^n(E,p) \cap B_s(p)$ such that $\frac{1}{c}(u_{ij}-p) = v_{ij}-p$. Thus, we get

$$\begin{aligned} x &= p + \sum_{i=1}^{m} \sum_{j=1}^{\infty} \alpha_{ij}(u_{ij}-p) = p + \sum_{i=1}^{m} \sum_{j=1}^{\infty} c\alpha_{ij}(v_{ij}-p) \\ &\in \mathrm{GS}(\mathrm{GS}^n(E,p) \cap B_s(p), p), \end{aligned}$$

which implies that $\mathrm{GS}(\mathrm{GS}^n(E,p) \cap B_t(p), p) \subset \mathrm{GS}(\mathrm{GS}^n(E,p) \cap B_s(p), p)$.

The reverse inclusion is obvious, since $B_s(p) \subset B_t(p)$. □

Proposition 4.15 guarantees the logical soundness of Definition 4.14 (i).

We now generalize Lemma 4.10 and formula (4.10) in the following lemma. Indeed, we prove that the function $T_{-q} \circ F_n \circ T_p : \mathrm{GS}^n(E_1,p)-p \to M_a$ preserves the inner product. This property is important in proving the following theorems as a necessary condition for F_n to be a d_a-isometry.

Lemma 4.16. *Let E_1 be a bounded subset of M_a that is d_a-isometric to a subset E_2 of M_a via a surjective d_a-isometry $f : E_1 \to E_2$. Assume that p and q are elements of E_1 and E_2, which satisfy $q = f(p)$. If $n \in \mathbb{N}$, then*

$$\big\langle (T_{-q} \circ F_n \circ T_p)(u-p),\, (T_{-q} \circ F_n \circ T_p)(v-p) \big\rangle_a = \langle u-p, v-p \rangle_a$$

for all $u, v \in \mathrm{GS}^n(E_1, p)$.

Proof. Our assertion for $n = 1$ was already proved in (4.10). Considering Proposition 4.15, assume that r is a positive real number satisfying $E_1 \subset B_r(p)$. Now we assume that the assertion is true for some $n \in \mathbb{N}$. Let u, v be arbitrary elements of $\mathrm{GS}^{n+1}(E_1, p)$. Then there exist some $m_1, m_2 \in \mathbb{N}$, $x_{ij}, y_{k\ell} \in \mathrm{GS}^n(E_1, p) \cap B_r(p)$, and some $\alpha_{ij}, \beta_{k\ell} \in \mathbb{R}$ such that

$$u - p = \sum_{i=1}^{m_1} \sum_{j=1}^{\infty} \alpha_{ij}(x_{ij} - p) \in M_a \quad \text{and} \quad v - p = \sum_{k=1}^{m_2} \sum_{\ell=1}^{\infty} \beta_{k\ell}(y_{k\ell} - p) \in M_a.$$

Using Definition 4.14 (ii) and our assumption, we get

$$\begin{aligned}
&\big\langle (T_{-q} \circ F_{n+1} \circ T_p)(u-p),\, (T_{-q} \circ F_{n+1} \circ T_p)(v-p) \big\rangle_a \\
&= \left\langle \sum_{i=1}^{m_1} \sum_{j=1}^{\infty} \alpha_{ij}(T_{-q} \circ F_n \circ T_p)(x_{ij}-p),\, \sum_{k=1}^{m_2} \sum_{\ell=1}^{\infty} \beta_{k\ell}(T_{-q} \circ F_n \circ T_p)(y_{k\ell}-p) \right\rangle_a \\
&= \sum_{i=1}^{m_1} \sum_{k=1}^{m_2} \sum_{j=1}^{\infty} \alpha_{ij} \sum_{\ell=1}^{\infty} \beta_{k\ell} \big\langle (T_{-q} \circ F_n \circ T_p)(x_{ij}-p),\, (T_{-q} \circ F_n \circ T_p)(y_{k\ell}-p) \big\rangle_a \\
&= \sum_{i=1}^{m_1} \sum_{k=1}^{m_2} \sum_{j=1}^{\infty} \alpha_{ij} \sum_{\ell=1}^{\infty} \beta_{k\ell} \langle x_{ij} - p, y_{k\ell} - p \rangle_a \\
&= \left\langle \sum_{i=1}^{m_1} \sum_{j=1}^{\infty} \alpha_{ij}(x_{ij} - p),\, \sum_{k=1}^{m_2} \sum_{\ell=1}^{\infty} \beta_{k\ell}(y_{k\ell} - p) \right\rangle_a \\
&= \langle u - p, v - p \rangle_a
\end{aligned}$$

for all $u, v \in \mathrm{GS}^{n+1}(E_1, p)$. By mathematical induction, we may then conclude that our assertion is true for all $n \in \mathbb{N}$. □

When $n = 1$ and $p = p'$, the first assertion in (i) of the following lemma is self-evident, so we have used that fact several times before, omitting the proof. The assertion (iv) in the following lemma seems to be related in some way to Proposition 4.15.

Lemma 4.17. *Assume that E is a bounded subset of M_a, p and p' are elements of E, and $n \in \mathbb{N}$. Let r be a positive real number satisfying $E \subset B_r(p)$.*

(i) $\mathrm{GS}^n(E,p) - p'$ *is a real vector space.*

(ii) $\mathrm{GS}^n(E,p) \subset \mathrm{GS}^{n+1}(E,p)$.

(iii) $\mathrm{GS}^2(E,p) = \overline{\mathrm{GS}(E,p)}$, *where* $\overline{\mathrm{GS}(E,p)}$ *is the closure of* $\mathrm{GS}(E,p)$ *in* M_a.

(iv) $\Lambda_\beta(\mathrm{GS}^n(E,p)) = \Lambda_\beta(\mathrm{GS}^n(E,p) \cap B_r(p))$, *where* $\beta = \{\beta_i\}_{i \in \mathbb{N}}$ *is a complete orthonormal sequence in* M_a.

Proof. (i) This claim is a generalization of the argument presented between Definition 4.14 and Proposition 4.15. By using Definitions 4.7 and 4.14, we prove that $\mathrm{GS}(E,p)-p'$ is a real vector space. (We can prove similarly for the case of $n > 1$.) Given $x, y \in \mathrm{GS}(E,p)-p'$, we may choose some $m_1, m_2 \in \mathbb{N}$, some $u_{ij}, v_{ij} \in E$, and some $\alpha_{ij}, \beta_{ij} \in \mathbb{R}$ such that $x = (p-p') + \sum_{i=1}^{m_1}\sum_{j=1}^{\infty} \alpha_{ij}(u_{ij}-p) \in M_a$ and $y = (p-p') + \sum_{i=1}^{m_2}\sum_{j=1}^{\infty} \beta_{ij}(v_{ij}-p) \in M_a$. Since M_a is a real vector space and $\mathrm{GS}(E,p) - p$ is a subspace of M_a, it holds that $\alpha \sum_{i=1}^{m_1}\sum_{j=1}^{\infty} \alpha_{ij}(u_{ij}-p) + \beta \sum_{i=1}^{m_2}\sum_{j=1}^{\infty} \beta_{ij}(v_{ij}-p) \in M_a$ for all $\alpha, \beta \in \mathbb{R}$.

Moreover, we see that

$$\begin{aligned}\alpha x + \beta y = &\left(p + (1-\alpha-\beta)(p'-p) + \sum_{i=1}^{m_1}\sum_{j=1}^{\infty} \alpha\alpha_{ij}(u_{ij}-p)\right.\\ &\left. + \sum_{i=1}^{m_2}\sum_{j=1}^{\infty} \beta\beta_{ij}(v_{ij}-p)\right) - p'\\ \in\ & \mathrm{GS}(E,p) - p'\end{aligned}$$

for all $\alpha, \beta \in \mathbb{R}$. Hence, $\mathrm{GS}(E,p) - p'$ is a real vector space as a subspace of real vector space M_a.

(ii) Let r be a positive real number with $E \subset B_r(p)$. If $x \in \mathrm{GS}^n(E,p)$ for some $n \in \mathbb{N}$, then $x - p \in \mathrm{GS}^n(E,p) - p$. Since $\mathrm{GS}^n(E,p) - p$ is a real vector space by (i) and $B_r(p) - p = B_r(0)$, we can choose a (sufficiently small) real number $\mu \neq 0$ such that $\mu(x-p) \in (\mathrm{GS}^n(E,p) - p) \cap (B_r(p) - p)$. We notice that

$$\begin{aligned}&\big(\mathrm{GS}^n(E,p) - p\big) \cap \big(B_r(p) - p\big)\\ &= \big\{v - p \in M_a : v \in \mathrm{GS}^n(E,p) \cap B_r(p)\big\}.\end{aligned} \tag{4.11}$$

Thus, we see that $\mu(x-p) = v - p$ for some $v \in \mathrm{GS}^n(E,p) \cap B_r(p)$. Since $x = p + \frac{1}{\mu}(v-p)$, it holds that $x \in \mathrm{GS}^{n+1}(E,p)$. Therefore, we conclude that $\mathrm{GS}^n(E,p) \subset \mathrm{GS}^{n+1}(E,p)$ for every $n \in \mathbb{N}$.

(iii) Let x be an arbitrary element of $\overline{\mathrm{GS}(E,p)}$. Then there exists some sequence $\{x_n\}$ that converges to x, where $x_n \in \mathrm{GS}(E,p) \setminus \{x\}$ for all $n \in \mathbb{N}$. We now set $y_1 = x_1$ and $y_i = x_i - x_{i-1}$ for each integer $i \geq 2$. Then we have

$$x_n = \sum_{i=1}^{n} y_i,$$

where $y_i = (x_i - p) - (x_{i-1} - p) \in \mathrm{GS}(E,p) - p$ for $i \geq 2$. Since $\mathrm{GS}(E,p) - p$ is a real vector space and $B_r(p) - p = B_r(0)$, we can select a real number $\mu_i \neq 0$ such that

$$\mu_i y_i \in \mathrm{GS}(E,p) - p \quad \text{and} \quad \mu_i y_i \in B_r(p) - p$$

for every integer $i \geq 2$. Thus, it follows from (4.11) that

$$x_n = \sum_{i=1}^{n} y_i = y_1 + \sum_{i=2}^{n} \frac{1}{\mu_i}(\mu_i y_i) = x_1 + \sum_{i=2}^{n} \frac{1}{\mu_i}(v_i - p),$$

where $v_i \in \mathrm{GS}(E,p) \cap B_r(p)$ for $i \geq 2$. Since the sequence $\{x_n\}$ is assumed to converge to x, the sequence $\left\{x_1 + \sum\limits_{i=2}^{n} \frac{1}{\mu_i}(v_i - p)\right\}_n$ converges to x. Hence, we have

$$x_1 + \sum_{i=2}^{\infty} \frac{1}{\mu_i}(v_i - p) = \lim_{n\to\infty} x_n = x \in M_a. \tag{4.12}$$

(Since M_a is a Hausdorff space, x is the unique limit point of the sequence $\{x_n\}$.)

Furthermore, there exists a real number $\mu_1 \neq 0$ that satisfies $\mu_1(x_1 - p) \in \mathrm{GS}(E,p) - p$ and $\mu_1(x_1 - p) \in B_r(p) - p$, i.e., $\mu_1(x_1 - p) \in (\mathrm{GS}(E,p) - p) \cap (B_r(p) - p)$. Thus, there exists a $v_1 \in \mathrm{GS}(E,p) \cap B_r(p)$ such that $\mu_1(x_1 - p) = v_1 - p$ or $x_1 - p = \frac{1}{\mu_1}(v_1 - p)$. Therefore,

$$x = p + (x_1 - p) + \sum_{i=2}^{\infty} \frac{1}{\mu_i}(v_i - p) = p + \sum_{i=1}^{\infty} \frac{1}{\mu_i}(v_i - p), \tag{4.13}$$

where $v_i \in \mathrm{GS}(E,p) \cap B_r(p)$ for each $i \in \mathbb{N}$. On account of (4.12), it holds that $\sum\limits_{i=1}^{\infty} \frac{1}{\mu_i}(v_i - p) \in M_a$. Thus, by (4.13), we see that $x \in \mathrm{GS}^2(E,p)$, which implies that $\overline{\mathrm{GS}(E,p)} \subset \mathrm{GS}^2(E,p)$.

On the other hand, let y be an arbitrary element of $\mathrm{GS}^2(E,p)$. Then there are some $m \in \mathbb{N}$, some $v_{ij} \in \mathrm{GS}(E,p) \cap B_r(p)$, and some $\alpha_{ij} \in \mathbb{R}$ such that

$y = p + \sum_{i=1}^{m}\sum_{j=1}^{\infty} \alpha_{ij}(v_{ij} - p) \in M_a$. Let us define $y_n = p + \sum_{i=1}^{m}\sum_{j=1}^{n} \alpha_{ij}(v_{ij} - p)$ for every $n \in \mathbb{N}$. Since $v_{ij} - p \in \mathrm{GS}(E,p) - p$ for all i and j and $\mathrm{GS}(E,p) - p$ is a real vector space, we know that $y_n - p = \sum_{i=1}^{m}\sum_{j=1}^{n} \alpha_{ij}(v_{ij} - p) \in \mathrm{GS}(E,p) - p$ and hence, $y_n \in \mathrm{GS}(E,p)$ for all $n \in \mathbb{N}$. Since $\mathrm{GS}(E,p)$ is a Hausdorff space, y is the unique element to which the sequence $\{y_n\}_{n\in\mathbb{N}}$ is convergent. Thus, we see that

$$y = p + \sum_{i=1}^{m}\sum_{j=1}^{\infty} \alpha_{ij}(v_{ij} - p) = \lim_{n\to\infty} y_n \in \overline{\mathrm{GS}(E,p)},$$

which implies that $\mathrm{GS}^2(E,p) \subset \overline{\mathrm{GS}(E,p)}$.

(iv) Let $i \in \Lambda_\beta(\mathrm{GS}^n(E,p))$. By Definition 4.8 (ii), there exist $x \in \mathrm{GS}^n(E,p)$ and $\alpha \neq 0$ with $x + \alpha\beta_i \in \mathrm{GS}^n(E,p)$. Further, $x = p + \sum_{i=1}^{m}\sum_{j=1}^{\infty} \alpha_{ij}(u_{ij} - p)$ for some $m \in \mathbb{N}$, some $u_{ij} \in \mathrm{GS}^{n-1}(E,p) \cap B_r(p)$, and for some $\alpha_{ij} \in \mathbb{R}$. Since $\sum_{i=1}^{m}\sum_{j=1}^{\infty} \alpha_{ij}(u_{ij} - p) + \alpha\beta_i = x - p + \alpha\beta_i \in \mathrm{GS}^n(E,p) - p$, where $\mathrm{GS}^n(E,p) - p$ is a real vector space and $B_r(p) - p = B_r(0)$, it holds that

$$\mu\left(\sum_{i=1}^{m}\sum_{j=1}^{\infty} \alpha_{ij}(u_{ij} - p) + \alpha\beta_i\right) \in (\mathrm{GS}^n(E,p) - p) \cap (B_r(p) - p)$$

for any sufficiently small $\mu \neq 0$, or equivalently, it follows from (4.11) that

$$\left(p + \sum_{i=1}^{m}\sum_{j=1}^{\infty} \mu\alpha_{ij}(u_{ij} - p)\right) + \mu\alpha\beta_i \in \mathrm{GS}^n(E,p) \cap B_r(p). \tag{4.14}$$

On the other hand, since $\sum_{i=1}^{m}\sum_{j=1}^{\infty} \alpha_{ij}(u_{ij} - p) = x - p \in \mathrm{GS}^n(E,p) - p$, it holds that $\sum_{i=1}^{m}\sum_{j=1}^{\infty} \mu\alpha_{ij}(u_{ij} - p) \in (\mathrm{GS}^n(E,p) - p) \cap (B_r(p) - p)$ for any sufficiently small $\mu \neq 0$. Thus, it follows from (4.11) that $p + \sum_{i=1}^{m}\sum_{j=1}^{\infty} \mu\alpha_{ij}(u_{ij} - p) \in \mathrm{GS}^n(E,p) \cap B_r(p)$ for any sufficiently small $\mu \neq 0$. Hence, by Definition 4.8 (ii) and (4.14), it holds that $i \in \Lambda_\beta(\mathrm{GS}^n(E,p) \cap B_r(p))$, which implies that $\Lambda_\beta(\mathrm{GS}^n(E,p)) \subset \Lambda_\beta(\mathrm{GS}^n(E,p) \cap B_r(p))$. Obviously, the inverse inclusion is true. □

As we mentioned earlier, we will see that the second-order generalized span is the last step in this kind of domain extension.

Proposition 4.18. *If E is a bounded subset of M_a and $p \in E$, then*

$$E \subset \mathrm{GS}(E,p) \subset \overline{\mathrm{GS}(E,p)} = \mathrm{GS}^2(E,p) = \mathrm{GS}^n(E,p)$$

for any integer $n \geq 2$. Indeed, $\mathrm{GS}^n(E,p) - p$ is a real Hilbert space for $n \geq 2$.

Proof. (a) Since E is bounded, we can choose a real number $r > 0$ that satisfies $E \subset B_r(p)$. Assume that $x \in \mathrm{GS}^3(E,p)$. Then there exist some $m_0 \in \mathbb{N}$, some $u_{ij} \in \mathrm{GS}^2(E,p) \cap B_r(p)$, and some $\alpha_{ij} \in \mathbb{R}$ such that $x = p + \sum_{i=1}^{m_0}\sum_{j=1}^{\infty} \alpha_{ij}(u_{ij} - p) \in M_a$.

We define $x_m = p + \sum_{i=1}^{m_0}\sum_{j=1}^{m} \alpha_{ij}(u_{ij} - p)$ for each $m \in \mathbb{N}$. Since $u_{ij} \in \mathrm{GS}^2(E,p)$, there exist some $m_{ij} \in \mathbb{N}$, some $v_{ijk\ell} \in \mathrm{GS}(E,p) \cap B_r(p)$, and some $\beta_{ijk\ell} \in \mathbb{R}$ such that $u_{ij} = p + \sum_{k=1}^{m_{ij}}\sum_{\ell=1}^{\infty} \beta_{ijk\ell}(v_{ijk\ell} - p) \in M_a$. Hence, it holds that

$$x_m = p + \sum_{i=1}^{m_0}\sum_{j=1}^{m}\sum_{k=1}^{m_{ij}}\sum_{\ell=1}^{\infty} \alpha_{ij}\beta_{ijk\ell}\big(v_{ijk\ell} - p\big) \in M_a,$$

which implies that $x_m \in \mathrm{GS}^2(E,p)$ for all $m \in \mathbb{N}$. Thus, $\{x_m\}$ is a sequence in $\mathrm{GS}^2(E,p)$ that converges to x. Therefore, $x \in \mathrm{GS}^2(E,p)$ because $\mathrm{GS}^2(E,p)$ is closed. Thus, $\mathrm{GS}^3(E,p) \subset \mathrm{GS}^2(E,p)$. The inverse inclusion is of course true due to Lemma 4.17 (ii). We have proved that $\mathrm{GS}^2(E,p) = \mathrm{GS}^3(E,p)$.

(b) According to Definition 4.14, we have

$$\mathrm{GS}^n(E,p) = \mathrm{GS}\big(\mathrm{GS}^{n-1}(E,p) \cap B_r(p), p\big)$$

for all $n \in \mathbb{N}$. If we replace n with $n+1$ in the above definition, then

$$\mathrm{GS}^{n+1}(E,p) = \mathrm{GS}\big(\mathrm{GS}^{n}(E,p) \cap B_r(p), p\big)$$

for any $n \in \mathbb{N}$. Hence, if $\mathrm{GS}^{n-1}(E,p) = \mathrm{GS}^n(E,p)$ for some integer $n \geq 3$, then it follows from the last two equalities that

$$\mathrm{GS}^n(E,p) = \mathrm{GS}^{n+1}(E,p).$$

With the conclusion of mathematical induction we prove that $\mathrm{GS}^n(E,p) = \mathrm{GS}^2(E,p)$ for every integer $n \geq 2$.

(c) Moreover, when $n \geq 2$, $\mathrm{GS}^n(E,p)$ is complete as a closed subset of a real Hilbert space M_a (ref. Theorem 1.41 and Remark 4.5). Therefore, $\mathrm{GS}^n(E,p) - p$ is a real Hilbert space for $n \geq 2$. □

The following lemma is an extension of Lemma 4.9 for the second-order generalized span $\mathrm{GS}^2(E,p)$. Indeed, we prove that if $i \in \Lambda_\beta(\mathrm{GS}^2(E,p))$, then the second-order generalized span of E contains all the lines through $\mathrm{GS}(E,p)$ in the direction β_i.

Lemma 4.19. *Assume that a bounded subset E of M_a contains at least two elements, $p \in E$, and $\beta = \{\beta_i\}_{i\in\mathbb{N}}$ is a complete orthonormal sequence in M_a. If $i \in \Lambda_\beta(\mathrm{GS}^2(E,p))$ and $p' \in \mathrm{GS}(E,p)$, then $p' + \alpha_i\beta_i \in \mathrm{GS}^2(E,p)$ for any $\alpha_i \in \mathbb{R}$.*

Proof. Let r be a positive real number with $E \subset B_r(p)$. Assume that $i \in \Lambda_\beta(\mathrm{GS}^2(E,p))$. Considering Lemma 4.17 (iv) and Proposition 4.18, if we substitute $\mathrm{GS}^2(E,p) \cap B_r(p)$ for E in Lemma 4.9, then $p + \alpha_i\beta_i \in \mathrm{GS}^3(E,p) = \mathrm{GS}^2(E,p)$ for all $\alpha_i \in \mathbb{R}$. Thus, there are some $m \in \mathbb{N}$, some $w_{ij} \in \mathrm{GS}(E,p) \cap B_r(p)$, and some $\gamma_{ij} \in \mathbb{R}$ with $\sum_{i=1}^{m}\sum_{j=1}^{\infty} \gamma_{ij}(w_{ij}-p) \in M_a$ such that $p + \alpha_i\beta_i = p + \sum_{i=1}^{m}\sum_{j=1}^{\infty} \gamma_{ij}(w_{ij}-p)$, and hence, we have

$$\begin{aligned} p' + \alpha_i\beta_i &= p + \alpha_i\beta_i + (p'-p) \\ &= p + \sum_{i=1}^{m}\sum_{j=1}^{\infty} \gamma_{ij}(w_{ij}-p) + (p'-p). \end{aligned} \tag{4.15}$$

Because $p'-p$ belongs to $\mathrm{GS}(E,p)-p$, which is a real vector space by Lemma 4.17 (i), and $B_r(p) - p = B_r(0)$, we can choose some sufficiently small real number $\mu \neq 0$ such that

$$\mu(p'-p) \in \mathrm{GS}(E,p) - p \quad \text{and} \quad \mu(p'-p) \in B_r(p) - p. \tag{4.16}$$

Considering (4.11), (4.15) and (4.16), if we put $\mu(p'-p) = w - p$ with a $w \in \mathrm{GS}(E,p) \cap B_r(p)$, then we have

$$p' + \alpha_i\beta_i = p + \sum_{i=1}^{m}\sum_{j=1}^{\infty} \gamma_{ij}(w_{ij}-p) + \frac{1}{\mu}(w-p) \in \mathrm{GS}^2(E,p)$$

for all $\alpha_i \in \mathbb{R}$. □

4.5 Basic Cylinders and Basic Intervals

We define infinite-dimensional intervals by classifying them as degenerate basic cylinders, non-degenerate basic cylinders, and basic intervals as follows.

Definition 4.20. For any positive integer n, we define the infinite-dimensional interval by

$$J = \prod_{i=1}^{\infty} J_i, \quad \text{where} \quad J_i = \begin{cases} [0, p_{2i}] & (\text{for } i \in \Lambda_1), \\ [p_{1i}, p_{2i}] & (\text{for } i \in \Lambda_2), \\ [p_{1i}, 1] & (\text{for } i \in \Lambda_3), \\ \{p_{1i}\} & (\text{for } i \in \Lambda_4), \\ [0, 1] & (\text{otherwise}) \end{cases}$$

for some disjoint finite subsets $\Lambda_1, \Lambda_2, \Lambda_3$ of $\{1, 2, \ldots, n\}$ and $0 < p_{1i} < p_{2i} < 1$ for $i \in \Lambda_1 \cup \Lambda_2 \cup \Lambda_3$ and $0 \leq p_{1i} \leq 1$ for $i \in \Lambda_4$. If $\Lambda_4 = \emptyset$, then J is called a *non-degenerate basic cylinder.* When Λ_4 is a nonempty finite set, J is called a *degenerate basic cylinder.* If Λ_4 is an infinite set, then J will be called a *basic interval.*

Remark 4.21. Let J be an infinite-dimensional interval.

(i) In order for J to become a basic cylinder, Λ_4 must be a finite set.

(ii) We remark that $\Lambda_4 = \mathbb{N} \setminus \Lambda(J)$ and $\Lambda(J) = \mathbb{N} \setminus \Lambda_4$. That is, $\mathbb{N}$ is the disjoint union of $\Lambda(J)$ and Λ_4 (see Definition 4.8 (i) for $\Lambda(J)$).

(iii) If $p = (p_1, p_2, \ldots, p_i, \ldots)$ is an element of J, then $J_i = \{p_i\}$ for each $i \notin \Lambda(J)$.

We note that the basic cylinder or the basic interval J defined in Definition 4.20 can be expressed as

$$J = \left\{ \sum_{i=1}^{\infty} \alpha_i \left(\frac{1}{a_i} e_i \right) : \alpha_i \in a_i J_i \text{ for all } i \in \mathbb{N} \right\},$$

where J_i is the interval defined in Definition 4.20 and $\left\{\frac{1}{a_i} e_i\right\}_{i \in \mathbb{N}}$ is a complete orthonormal sequence in M_a.

Let $\beta = \{\beta_i\}_{i \in \mathbb{N}}$ be a complete orthonormal sequence in M_a. We now consider a d_a-isometry $U : M_a \to M_a$ defined by $U(x) := \sum_{i=1}^{\infty} \alpha_i \beta_i$ for all $x = \sum_{i=1}^{\infty} \alpha_i \frac{1}{a_i} e_i \in M_a$. The d_a-isometry U maps $\frac{1}{a_i} e_i$ to β_i for any $i \in \mathbb{N}$. Considering this example, we define the β-basic cylinder and the β-basic interval as follows.

Definition 4.22. Let $\beta = \{\beta_i\}_{i \in \mathbb{N}}$ be a complete orthonormal sequence in M_a, J_i the interval given in Definition 4.20, and let n be a positive integer. We define

$$J_\beta = \left\{ \sum_{i=1}^{\infty} \alpha_i \beta_i : \alpha_i \in a_i J_i \text{ for all } i \in \mathbb{N} \right\},$$

for some disjoint finite subsets $\Lambda_1, \Lambda_2, \Lambda_3$ of $\{1, 2, \ldots, n\}$ and $0 < p_{1i} < p_{2i} < 1$ for $i \in \Lambda_1 \cup \Lambda_2 \cup \Lambda_3$ and $0 \leq p_{1i} \leq 1$ for $i \in \Lambda_4$. If $\Lambda_4 = \emptyset$, then J_β is called a *non-degenerate β-basic cylinder.* When Λ_4 is a nonempty finite set, J_β is called a *degenerate β-basic cylinder.* If Λ_4 is an infinite set, then J_β will be called a *β-basic interval.*

We note that $J_\beta = U(J)$, where J is a basic cylinder or a basic interval defined in Definition 4.20 and J_β is a β-basic cylinder or a β-basic interval defined in Definition 4.22.

Using Definitions 4.20 and 4.22, Remark 4.21 (ii) is generalized to:

Remark 4.23. Let $\beta = \{\beta_i\}_{i \in \mathbb{N}}$ be a complete orthonormal sequence in M_a and let J_β be a translation of a β-basic cylinder or a translation of a β-basic interval.

(i) $\Lambda_\beta(J_\beta) = \mathbb{N} \backslash \Lambda_4$, where Λ_4 is given in Definitions 4.20 and 4.22 and $\Lambda_\beta(J_\beta)$ is defined in Definition 4.8 (ii).

(ii) If $p = \sum_{i=1}^{\infty} p_i \beta_i$ and $x = \sum_{i=1}^{\infty} x_i \beta_i$ are elements of J_β, then $p_i = x_i$ for each $i \notin \Lambda_\beta(J_\beta)$.

Proof. (i) If $i \in \Lambda_4$, then it follows from Definition 4.22 that

$$\langle x, \beta_i \rangle_a = \left\langle p + \sum_{j=1}^{\infty} \alpha_j \beta_j, \beta_i \right\rangle_a = p_i + \alpha_i \in p_i + a_i J_i = \{p_i + a_i p_{1i}\}$$

for all $x \in J_\beta$, where $p = \sum_{i=1}^{\infty} p_i \beta_i$ is a fixed element of M_a such that $J_\beta = T_p(J'_\beta)$ for some β-basic cylinder or β-basic interval J'_β. That is, $\langle x, \beta_i \rangle_a = p_i + a_i p_{1i}$ for all $x \in J_\beta$. If $i \in \Lambda_4$, then $\langle x + \alpha \beta_i, \beta_i \rangle_a = \langle x, \beta_i \rangle_a + \alpha = p_i + a_i p_{1i} + \alpha \neq p_i + a_i p_{1i}$ for all $x \in J_\beta$ and $\alpha \neq 0$, which implies that $x + \alpha \beta_i \notin J_\beta$. That is, in view of Definition 4.8 (ii), we conclude that $i \notin \Lambda_\beta(J_\beta)$.

We now assume that $i \notin \Lambda_\beta(J_\beta)$. Then by Definition 4.8 (ii), it holds that

$$x + \alpha \beta_i \notin J_\beta \tag{4.17}$$

for any $x \in J_\beta$ and $\alpha \neq 0$. Using Definition 4.22, we have

$$x + \alpha \beta_i = p + \sum_{j \notin \Lambda_4} \alpha_j \beta_j + \sum_{j \in \Lambda_4} a_j p_{1j} \beta_j + \alpha \beta_i \tag{4.18}$$

for all $x \in J_\beta$ and $\alpha \neq 0$ and for some $p \in M_a$. We assume on the contrary that $i \notin \Lambda_4$. Since $\alpha_i \in a_i J_i$ and $a_i J_i$ is an interval with nonzero Euclidean length for

$i \notin \Lambda_4$, there exists an $\alpha \neq 0$ that satisfies $\alpha_i + \alpha \in a_i J_i$. In view of Definition 4.22 and (4.18), it holds that

$$x + \alpha\beta_i = p + \sum_{j \notin \Lambda_4 \cup \{i\}} \alpha_j \beta_j + (\alpha_i + \alpha)\beta_i + \sum_{j \in \Lambda_4} a_j p_{1j} \beta_j \in J_\beta$$

for some $x \in J_\beta$ and $\alpha \neq 0$, which is contrary to (4.17). Therefore, we conclude that if $i \notin \Lambda_\beta(J_\beta)$, then $i \in \Lambda_4$.

(ii) If $i \notin \Lambda_\beta(J_\beta)$ then $i \in \Lambda_4$ by (i). Furthermore, if $p = \sum_{i=1}^{\infty} p_i \beta_i \in J_\beta$ and $x = \sum_{i=1}^{\infty} x_i \beta_i \in J_\beta$, then it follows from Definitions 4.20 and 4.22 that $p_i = p_i' + a_i p_{1i} = x_i$, where $p' = \sum_{i=1}^{\infty} p_i' \beta_i$ is a fixed element of M_a such that $J_\beta = T_{p'}(J_\beta')$ for some β-basic cylinder or β-basic interval J_β'. □

Theorem 4.24. *Let $\beta = \{\beta_i\}_{i \in \mathbb{N}}$ be a complete orthonormal sequence in M_a and let J_β be either a translation of a β-basic cylinder or a translation of a β-basic interval and $p \in J_\beta$. Then*

$$\mathrm{GS}(J_\beta, p) = \left\{ p + \sum_{i \in \Lambda_\beta(J_\beta)} \alpha_i \beta_i \in M_a : \alpha_i \in \mathbb{R} \textit{ for all } i \in \Lambda_\beta(J_\beta) \right\}.$$

Proof. Assume that x is an arbitrary element of $\mathrm{GS}(J_\beta, p)$. By Definition 4.7, we have

$$x - p = \sum_{i=1}^{m} \sum_{j=1}^{\infty} \varepsilon_{ij}(x_{ij} - p) \in M_a$$

for some $m \in \mathbb{N}$, $\varepsilon_{ij} \in \mathbb{R}$, and $x_{ij} \in J_\beta$. Furthermore, since $x_{ij}, p \in J_\beta$, by Definition 4.22, we get

$$x_{ij} = p' + \sum_{k=1}^{\infty} \gamma_k \beta_k = p' + \sum_{k \in \mathbb{N} \setminus \Lambda_4} \gamma_k \beta_k + \sum_{k \in \Lambda_4} a_k p_{1k} \beta_k$$

and

$$p = p' + \sum_{k=1}^{\infty} \delta_k \beta_k = p' + \sum_{k \in \mathbb{N} \setminus \Lambda_4} \delta_k \beta_k + \sum_{k \in \Lambda_4} a_k p_{1k} \beta_k$$

for some $p' \in M_a$ and $\gamma_k, \delta_k \in a_k J_k$, where $J_\beta = T_{p'}(J_\beta')$ for some β-basic cylinder or β-basic interval J_β'.

Since $\{\beta_i\}_{i\in\mathbb{N}}$ is a complete orthonormal sequence in M_a, it follows from Definition 4.22 and Remark 4.23 (i) that

$$x - p = \sum_{i=1}^{m}\sum_{j=1}^{\infty} \varepsilon_{ij}(x_{ij} - p) = \sum_{i=1}^{m}\sum_{j=1}^{\infty} \varepsilon_{ij} \sum_{k\in\mathbb{N}\setminus\Lambda_4} (\gamma_k - \delta_k)\beta_k$$
$$= \sum_{k\in\mathbb{N}\setminus\Lambda_4} \omega_k\beta_k = \sum_{i\in\Lambda_\beta(J'_\beta)} \omega_i\beta_i = \sum_{i\in\Lambda_\beta(J_\beta)} \omega_i\beta_i$$

for some real numbers ω_i. We note that $\Lambda_\beta(J_\beta) = \Lambda_\beta(J'_\beta)$. Since $x \in \mathrm{GS}(J_\beta, p) \subset M_a$, it holds that

$$x = p + \sum_{i\in\Lambda_\beta(J_\beta)} \omega_i\beta_i \; (\in M_a)$$
$$\in \left\{ p + \sum_{i\in\Lambda_\beta(J_\beta)} \alpha_i\beta_i \in M_a : \alpha_i \in \mathbb{R} \text{ for all } i \in \Lambda_\beta(J_\beta) \right\},$$

which implies that

$$\mathrm{GS}(J_\beta, p) \subset \left\{ p + \sum_{i\in\Lambda_\beta(J_\beta)} \alpha_i\beta_i \in M_a : \alpha_i \in \mathbb{R} \text{ for all } i \in \Lambda_\beta(J_\beta) \right\}.$$

It remains to prove the reverse inclusion. According to the structure of J_β given in Definition 4.22, for each $i \in \Lambda_\beta(J_\beta)$, there exists a real number $\gamma_i \neq 0$ such that $p + \gamma_i\beta_i \in J_\beta$. In other words, for each $i \in \Lambda_\beta(J_\beta)$, there exists a $u_i \in J_\beta$ such that $\gamma_i\beta_i = u_i - p$. Thus, if we assume that

$$p + \sum_{i\in\Lambda_\beta(J_\beta)} \alpha_i\beta_i \in M_a$$

for some $\alpha_i \in \mathbb{R}$, then

$$p + \sum_{i\in\Lambda_\beta(J_\beta)} \alpha_i\beta_i = p + \sum_{i\in\Lambda_\beta(J_\beta)} \frac{\alpha_i}{\gamma_i}(\gamma_i\beta_i) = p + \sum_{i\in\Lambda_\beta(J_\beta)} \frac{\alpha_i}{\gamma_i}(u_i - p)$$
$$\in \mathrm{GS}(J_\beta, p),$$

since $u_i \in J_\beta$ for all $i \in \Lambda_\beta(J_\beta)$, which implies that

$$\mathrm{GS}(J_\beta, p) \supset \left\{ p + \sum_{i\in\Lambda_\beta(J_\beta)} \alpha_i\beta_i \in M_a : \alpha_i \in \mathbb{R} \text{ for all } i \in \Lambda_\beta(J_\beta) \right\}.$$

We end the proof in this way. □

In the following theorem, we introduce an interesting inclusion property of the second-order generalized span.

Theorem 4.25. *Assume that H is a closed subspace of M_a. Let $\beta = \{\beta_i\}_{i \in \mathbb{N}}$ and $\{\beta_i\}_{i \in \Lambda}$ be complete orthonormal sequences in the Hilbert spaces M_a and H, respectively, and let J_β be either a translation of a β-basic cylinder or a translation of a β-basic interval that satisfies $\Lambda_\beta(J_\beta) \subset \Lambda$. Then $J_\beta - p \subset \mathrm{GS}(J_\beta, p) - p \subset H$ for any $p \in J_\beta$.*

Proof. Assume that $p \in J_\beta$ and $x \in \mathrm{GS}(J_\beta, p) - p$. Then, according to Theorem 4.24, there exist some real numbers α_i that satisfy

$$x = \sum_{i \in \Lambda_\beta(J_\beta)} \alpha_i \beta_i \in M_a. \tag{4.19}$$

Assume that $i \in \Lambda_\beta(J_\beta)$. Then $i \in \Lambda$. Since H is a real vector space and $\beta_i \in H$, it holds that

$$\alpha_i \beta_i \in H \tag{4.20}$$

for all $i \in \Lambda_\beta(J_\beta)$. Now we define

$$x_n := \sum_{i \in \Lambda_n} \alpha_i \beta_i \in H$$

for any $n \in \mathbb{N}$, where we set $\Lambda_n = \{i \in \Lambda_\beta(J_\beta) : i < n\}$. Since H is assumed to be closed, it follows from (4.19) that

$$x = \lim_{n \to \infty} x_n \in H,$$

which implies that $\mathrm{GS}(J_\beta, p) - p \subset H$. □

Since in some ways index sets have some properties of dimensions in vector space, the following theorem may seem obvious.

Theorem 4.26. *Let $\beta = \{\beta_i\}_{i \in \mathbb{N}}$ be a complete orthonormal sequence in M_a. Assume that a bounded subset E of M_a contains at least two elements and $p \in E$. Then, $\Lambda_\beta(\mathrm{GS}^2(E, p)) = \mathbb{N}$ if and only if $\mathrm{GS}^2(E, p) = M_a$.*

Proof. Let x be an arbitrary element of M_a. Then there exist some real numbers α_i such that

$$x = \sum_{i=1}^{\infty} \alpha_i \beta_i \in M_a. \tag{4.21}$$

If $\Lambda_\beta(\mathrm{GS}^2(E,p)) = \mathbb{N}$, then it follows from Lemma 4.19 that

$$p + \alpha_i\beta_i \in \mathrm{GS}^2(E,p)$$

for all $i \in \mathbb{N}$. In other words,

$$\alpha_i\beta_i \in \mathrm{GS}^2(E,p) - p$$

for all $i \in \mathbb{N}$.

By Lemma 4.17 (i), we get

$$x_n := \sum_{i=1}^{n} \alpha_i\beta_i \in \mathrm{GS}^2(E,p) - p$$

for any $n \in \mathbb{N}$. Due to Lemma 4.17 (iii) and (4.21), we further obtain

$$x = \sum_{i=1}^{\infty} \alpha_i\beta_i = \lim_{n\to\infty} x_n \in \mathrm{GS}^2(E,p) - p,$$

which implies that $M_a \subset \mathrm{GS}^2(E,p) - p$, or equivalently, $M_a \subset \mathrm{GS}^2(E,p)$.

The reverse inclusion is trivial. □

4.6 Second-Order Extension of Isometries

It was proved in Theorem 4.13 that the domain of a d_a-isometry $f : E_1 \to E_2$ can be extended to the first-order generalized span $\mathrm{GS}(E_1,p)$ whenever E_1 is a nonempty bounded subset of M_a, whether degenerate or non-degenerate.

Now we generalize Theorem 4.13 in the following theorem. More precisely, we prove that the domain of f can be extended to its second-order generalized span $\mathrm{GS}^2(E_1,p)$. We note that $\mathrm{GS}^2(E_1,p) = \overline{\mathrm{GS}(E_1,p)}$ by Lemma 4.17 (iii). Therefore, Theorem 4.27 is a further generalization of [6, Theorem 2.2].

In the proof, we use the fact that $\mathrm{GS}^n(E_1,p) - p$ is a real vector space.

Theorem 4.27. *Let E_1 be a bounded subset of M_a that is d_a-isometric to a subset E_2 of M_a via a surjective d_a-isometry $f : E_1 \to E_2$. Assume that p and q are elements of E_1 and E_2, which satisfy $q = f(p)$. The function $F_2 : \mathrm{GS}^2(E_1,p) \to M_a$ is a d_a-isometry and the function $T_{-q} \circ F_2 \circ T_p : \mathrm{GS}^2(E_1,p) - p \to M_a$ is linear. In particular, F_2 is an extension of F.*

Proof. (a) Suppose r is a positive real number satisfying $E_1 \subset B_r(p)$. Referring to the changes presented in the table below and following the first part of proof of Theorem 4.13, we can easily prove that F_2 is a d_a-isometry.

Theorem 4.13:	E_1	$\mathrm{GS}(E_1,p)$	f	F
Here:	$\mathrm{GS}(E_1,p)\cap B_r(p)$	$\mathrm{GS}^2(E_1,p)$	F	F_2

Theorem 4.13:	Definition 4.11	Lemma 4.10
Here:	Definition 4.14	Lemma 4.16

(b) Referring to the changes presented in the table below and following (d) of the proof of Theorem 4.13, we can prove the linearity of $T_{-q}\circ F_n\circ T_p$: $\mathrm{GS}^n(E_1,p)-p\to M_a$ in a more general setting for $n\geq 2$.

Theorem 4.13:	$\mathrm{GS}(E_1,p)$	F	(4.10)
Here:	$\mathrm{GS}^n(E_1,p)$	F_n	Lemma 4.16

(c) According to Definition 4.14 (i), for any $m\in\mathbb{N}$, $x_{ij}\in\mathrm{GS}(E_1,p)\cap B_r(p)$, and any $\alpha_{ij}\in\mathbb{R}$ with $\sum_{i=1}^{m}\sum_{j=1}^{\infty}\alpha_{ij}(x_{ij}-p)\in M_a$, there exists a $u\in\mathrm{GS}^2(E_1,p)$ satisfying

$$u-p=\sum_{i=1}^{m}\sum_{j=1}^{\infty}\alpha_{ij}(x_{ij}-p)\in M_a. \tag{4.22}$$

Due to Definition 4.14 (ii), we further have

$$(T_{-q}\circ F_2\circ T_p)(u-p)=\sum_{i=1}^{m}\sum_{j=1}^{\infty}\alpha_{ij}(T_{-q}\circ F\circ T_p)(x_{ij}-p). \tag{4.23}$$

If we set $\alpha_{11}=1$, $\alpha_{ij}=0$ for each $(i,j)\neq(1,1)$, and $x_{11}=x$ in (4.22) and (4.23) to see

$$(T_{-q}\circ F_2\circ T_p)(x-p)=(T_{-q}\circ F\circ T_p)(x-p) \tag{4.24}$$

for all $x\in\mathrm{GS}(E_1,p)\cap B_r(p)$.

Let w be an arbitrary element of $\mathrm{GS}(E_1,p)$. Then, $w-p\in\mathrm{GS}(E_1,p)-p$. Since $\mathrm{GS}(E_1,p)-p$ is a real vector space and $B_r(p)-p=B_r(0)$, there exists a real number $\mu\neq 0$ such that

$$\mu(w-p)\in(\mathrm{GS}(E_1,p)-p)\cap(B_r(p)-p).$$

Thus, by (4.11), we can choose a $v\in\mathrm{GS}(E_1,p)\cap B_r(p)$ such that $\mu(w-p)=v-p$. Since both $T_{-q}\circ F_2\circ T_p$ and $T_{-q}\circ F\circ T_p$ are linear and $\mathrm{GS}(E_1,p)\subset\mathrm{GS}^2(E_1,p)$, it follows from (4.24) that

$$\begin{aligned}\mu(T_{-q}\circ F_2\circ T_p)(w-p) &= (T_{-q}\circ F_2\circ T_p)(\mu(w-p))\\ &= (T_{-q}\circ F_2\circ T_p)(v-p)\\ &= (T_{-q}\circ F\circ T_p)(v-p)\\ &= (T_{-q}\circ F\circ T_p)(\mu(w-p))\\ &= \mu(T_{-q}\circ F\circ T_p)(w-p).\end{aligned}$$

Therefore, it follows that $(T_{-q}\circ F_2\circ T_p)(w-p) = (T_{-q}\circ F\circ T_p)(w-p)$ for all $w\in \mathrm{GS}(E_1,p)$, i.e., $F_2(w) = F(w)$ for all $w\in \mathrm{GS}(E_1,p)$. In other words, F_2 is an extension of F. Also, because of Theorem 4.13, we see that F_2 is obviously an extension of f. □

On account of Proposition 4.18, it holds that

$$\mathrm{GS}^2(E_1,p) = \cdots = \mathrm{GS}^{n-1}(E_1,p) = \mathrm{GS}^n(E_1,p)$$

for every integer $n\geq 3$. According to this formula, the assertion of the following theorem seems obvious, but since the proof is not long, we introduce the proof here.

Theorem 4.28. *Let E_1 be a bounded subset of M_a that is d_a-isometric to a subset E_2 of M_a via a surjective d_a-isometry $f: E_1\to E_2$. Assume that p and q are elements of E_1 and E_2, which satisfy $q = f(p)$. Then F_n is identically the same as F_2 for any integer $n\geq 3$, where F_2 and F_n are defined in Definition 4.14.*

Proof. Let r be a fixed positive real number satisfying $E_1\subset B_r(p)$. We assume that $F_2\equiv F_3\equiv\cdots\equiv F_{n-1}$ on $\mathrm{GS}^2(E_1,p)$. Let x be an arbitrary element of $\mathrm{GS}^n(E_1,p)$. Then, in view of (4.11), there exists a real number $\mu\neq 0$ and an element u of $\mathrm{GS}^n(E_1,p)\cap B_r(p)$ such that

$$u-p = \mu(x-p)\in (\mathrm{GS}^n(E_1,p)-p)\cap(B_r(p)-p).$$

If we put $\alpha_{11}=1$, $\alpha_{ij}=0$ for all $(i,j)\neq(1,1)$, and $x_{11}=v$ in Definition 4.14 (ii), then we get

$$(T_{-q}\circ F_n\circ T_p)(v-p) = (T_{-q}\circ F_{n-1}\circ T_p)(v-p) \tag{4.25}$$

for all $v\in \mathrm{GS}^{n-1}(E_1,p)\cap B_r(p) = \mathrm{GS}^n(E_1,p)\cap B_r(p)$. We note by Proposition 4.18 that $\mathrm{GS}^n(E_1,p) = \mathrm{GS}^{n-1}(E_1,p) = \cdots = \mathrm{GS}^2(E_1,p)$.

Since $T_{-q} \circ F_n \circ T_p$ is linear by (b) in the proof of Theorem 4.27, it follows from (4.25) that

$$\begin{aligned}\mu(T_{-q} \circ F_n \circ T_p)(x-p) &= (T_{-q} \circ F_n \circ T_p)(u-p) \\ &= (T_{-q} \circ F_{n-1} \circ T_p)(u-p) \\ &= (T_{-q} \circ F_2 \circ T_p)(u-p) \\ &= \mu(T_{-q} \circ F_2 \circ T_p)(x-p),\end{aligned}$$

i.e., $F_n(x) = F_2(x)$ for every $x \in \mathrm{GS}^n(E_1,p) = \mathrm{GS}^2(E_1,p)$. By mathematical induction, we conclude that F_n is identically the same as F_2 for every integer $n \geq 3$. □

In the following two remarks, let $\beta = \{\beta_i\}_{i\in\mathbb{N}}$ be a complete orthonormal sequence in M_a and let J_β be either a translation of a β-basic cylinder or a translation of a β-basic interval and $p \in J_\beta$. Due to Definition 4.22, Remark 4.23, and Theorem 4.24, $\mathrm{GS}(J_\beta,p)$ is a closed subset of M_a.

Remark 4.29. $\mathrm{GS}(J_\beta,p)$ is a closed subset of M_a.

Proof. Assume that $p = \sum_{i=1}^{\infty} p_i\beta_i$ is a fixed element of J_β, where J_β is a translation of a β-basic cylinder or a translation of a β-basic interval. In view of Definition 4.7 and Remark 4.23 (ii), we note that $x_i = p_i$ for each $x = \sum_{i=1}^{\infty} x_i\beta_i \in \mathrm{GS}(J_\beta,p)$ and each $i \notin \Lambda_\beta(J_\beta)$.

Assume that $\{z_n\}_{n\in\mathbb{N}}$ is a sequence of elements in $\mathrm{GS}(J_\beta,p)$, which converges to an element $z = \sum_{i=1}^{\infty} z_i\beta_i$ of M_a, where we set $z_n = \sum_{i=1}^{\infty} z_{ni}\beta_i$ for any $n \in \mathbb{N}$. Since $z_n \in \mathrm{GS}(J_\beta,p)$ for every $n \in \mathbb{N}$, the previous argument implies that $z_{ni} = p_i$ for each $i \notin \Lambda_\beta(J_\beta)$. Thus, we conclude that $z_i = p_i$ for each $i \notin \Lambda_\beta(J_\beta)$. This fact, together with Theorem 4.24, implies that $z \in \mathrm{GS}(J_\beta,p)$. Therefore, we conclude that $\mathrm{GS}(J_\beta,p)$ is a closed subset of M_a. □

On account of Theorem 4.24, we note that $\Lambda_\beta(J_\beta) = \Lambda_\beta(\mathrm{GS}(J_\beta,p))$.

Remark 4.30. $\mathrm{GS}^2(J_\beta,p) = \mathrm{GS}(J_\beta,p)$.

Proof. We note that $\mathrm{GS}(J_\beta,p)$ is a closed subset of M_a by Remark 4.29. Referring to the changes presented in the table below

Proposition 4.18:	$\mathrm{GS}(E,p) \cap B_r(p)$	$\mathrm{GS}^2(E,p)$	$\mathrm{GS}^3(E,p)$	x	x_m
Here:	J_β	$\mathrm{GS}(J_\beta,p)$	$\mathrm{GS}^2(J_\beta,p)$	u	u_m

and following the part (a) in the proof of Proposition 4.18, we can easily show that $\mathrm{GS}^2(J_\beta, p) = \mathrm{GS}(J_\beta, p)$. □

Hence, by Theorem 4.24 with $\beta = \left\{\frac{1}{a_i} e_i\right\}_{i\in\mathbb{N}}$ and Remark 4.30, we have

$$\begin{aligned} u - p &= \sum_{i=1}^{\infty} \left\langle u - p, \frac{1}{a_i} e_i \right\rangle_a \frac{1}{a_i} e_i \\ &= \sum_{i=1}^{\infty} a_i (u_i - p_i) \frac{1}{a_i} e_i \\ &= \sum_{i\in\Lambda(J)} a_i (u_i - p_i) \frac{1}{a_i} e_i \\ &= \sum_{i\in\Lambda(J)} \left\langle u - p, \frac{1}{a_i} e_i \right\rangle_a \frac{1}{a_i} e_i \end{aligned} \tag{4.26}$$

for all $u \in \mathrm{GS}^2(J, p) = \mathrm{GS}^n(J, p)$, where $n \in \mathbb{N}$.

Using a similar approach to the proof of [10, Theorem 2.4], we can apply Lemma 4.16 to prove the following theorem.

Theorem 4.31. *Assume that J is either a translation of a basic cylinder or a translation of a basic interval, K is a subset of M_a, and that there exists a surjective d_a-isometry $f : J \to K$. Suppose p is an element of J and q is an element of K with $q = f(p)$. For any $n \in \mathbb{N}$, the d_a-isometry $F_n : \mathrm{GS}^n(J, p) \to M_a$ given in Definition 4.14 satisfies*

$$(T_{-q} \circ F_n \circ T_p)(u - p) = \sum_{i\in\Lambda(J)} \left\langle u - p, \frac{1}{a_i} e_i \right\rangle_a \frac{1}{a_i} (T_{-q} \circ F_n \circ T_p)(e_i)$$

for all $u \in \mathrm{GS}^n(J, p)$.

Proof. Since $p + e_i \in \mathrm{GS}^n(J, p)$ for each $i \in \Lambda(J)$, it follows from Lemma 4.16 that

$$\begin{aligned}
&\Big\langle (T_{-q} \circ F_n \circ T_p)(u-p) - \sum_{i \in \Lambda(J)} \Big\langle u-p, \frac{1}{a_i} e_i \Big\rangle_a \frac{1}{a_i} (T_{-q} \circ F_n \circ T_p)(e_i), \\
&\quad (T_{-q} \circ F_n \circ T_p)(u-p) - \sum_{j \in \Lambda(J)} \Big\langle u-p, \frac{1}{a_j} e_j \Big\rangle_a \frac{1}{a_j} (T_{-q} \circ F_n \circ T_p)(e_j) \Big\rangle_a \\
&= \langle (T_{-q} \circ F_n \circ T_p)(u-p), (T_{-q} \circ F_n \circ T_p)(u-p) \rangle_a \\
&\quad - \sum_{j \in \Lambda(J)} \Big\langle u-p, \frac{1}{a_j} e_j \Big\rangle_a \frac{1}{a_j} \langle (T_{-q} \circ F_n \circ T_p)(u-p), (T_{-q} \circ F_n \circ T_p)(e_j) \rangle_a \\
&\quad - \sum_{i \in \Lambda(J)} \Big\langle u-p, \frac{1}{a_i} e_i \Big\rangle_a \frac{1}{a_i} \langle (T_{-q} \circ F_n \circ T_p)(e_i), (T_{-q} \circ F_n \circ T_p)(u-p) \rangle_a \\
&\quad + \sum_{i \in \Lambda(J)} \sum_{j \in \Lambda(J)} \Big\langle u-p, \frac{1}{a_i} e_i \Big\rangle_a \Big\langle u-p, \frac{1}{a_j} e_j \Big\rangle_a \times \\
&\qquad\qquad \times \frac{1}{a_i a_j} \langle (T_{-q} \circ F_n \circ T_p)(e_i), (T_{-q} \circ F_n \circ T_p)(e_j) \rangle_a \\
&= \langle u-p, u-p \rangle_a - \sum_{j \in \Lambda(J)} \Big\langle u-p, \frac{1}{a_j} e_j \Big\rangle_a \Big\langle u-p, \frac{1}{a_j} e_j \Big\rangle_a \\
&\quad - \sum_{i \in \Lambda(J)} \Big\langle u-p, \frac{1}{a_i} e_i \Big\rangle_a \Big\langle \frac{1}{a_i} e_i, u-p \Big\rangle_a \\
&\quad + \sum_{i \in \Lambda(J)} \sum_{j \in \Lambda(J)} \Big\langle u-p, \frac{1}{a_i} e_i \Big\rangle_a \Big\langle u-p, \frac{1}{a_j} e_j \Big\rangle_a \Big\langle \frac{1}{a_i} e_i, \frac{1}{a_j} e_j \Big\rangle_a \\
&= \langle u-p, u-p \rangle_a - \sum_{j \in \Lambda(J)} \Big\langle u-p, \frac{1}{a_j} e_j \Big\rangle_a \Big\langle u-p, \frac{1}{a_j} e_j \Big\rangle_a
\end{aligned}$$

for all $u \in \mathrm{GS}^n(J, p)$, since $\{\frac{1}{a_i} e_i\}_{i \in \mathbb{N}}$ is an orthonormal sequence in M_a.

Furthermore, we note that each $u \in \mathrm{GS}^n(J, p)$ has the expression given in (4.26). Hence, if we replace $u - p$ in the previous equalities with the expression (4.26), then we have

$$\Big\| (T_{-q} \circ F_n \circ T_p)(u-p) - \sum_{i \in \Lambda(J)} \Big\langle u-p, \frac{1}{a_i} e_i \Big\rangle_a \frac{1}{a_i} (T_{-q} \circ F_n \circ T_p)(e_i) \Big\|_a^2 = 0$$

for all $u \in \mathrm{GS}^n(J, p)$, which implies the validity of our assertion. $\square$

According to the following theorem, the image of the first-order generalized span of E_1 with respect to p under the d_a-isometry F is just the first-order generalized span of $F(E_1)$ with respect to $F(p)$. This assertion holds also for the second-order generalized span and F_2. According to Proposition 4.18 and Theorem 4.28, the argument of the following theorem only makes sense when $n = 1$ or 2.

Theorem 4.32. *Assume that E_1 and E_2 are bounded subsets of M_a that are d_a-isometric to each other via a surjective d_a-isometry $f : E_1 \to E_2$. Suppose p is an element of E_1 and q is an element of E_2 with $q = f(p)$. If $F_n : \mathrm{GS}^n(E_1, p) \to M_a$ is the extension of f defined in Definition 4.14, then $\mathrm{GS}^n(E_2, q) = F_n(\mathrm{GS}^n(E_1, p))$ for every $n \in \mathbb{N}$.*

Proof. (a) First, we prove that our assertion is true for $n = 1$, i.e., we prove that $\mathrm{GS}(E_2, q) = F(\mathrm{GS}(E_1, p))$. Let r be a fixed positive real number satisfying $E_1 \subset B_r(p)$.

(b) Due to Definition 4.7, for any $y \in F(\mathrm{GS}(E_1, p))$, there exists an element $x \in \mathrm{GS}(E_1, p)$ with

$$y = F(x) = F\left(p + \sum_{i=1}^{m} \sum_{j=1}^{\infty} \alpha_{ij}(u_{ij} - p)\right)$$

for some $m \in \mathbb{N}$, $u_{ij} \in E_1 \cap B_r(p)$, and some $\alpha_{ij} \in \mathbb{R}$ which satisfy the condition $x = p + \sum\limits_{i=1}^{m} \sum\limits_{j=1}^{\infty} \alpha_{ij}(u_{ij} - p) \in M_a$.

On the other hand, by Definition 4.11, we have

$$(T_{-q} \circ F \circ T_p)\left(\sum_{i=1}^{m} \sum_{j=1}^{\infty} \alpha_{ij}(u_{ij} - p)\right) = \sum_{i=1}^{m} \sum_{j=1}^{\infty} \alpha_{ij}(T_{-q} \circ f \circ T_p)(u_{ij} - p)$$

which is equivalent to

$$F(x) - q = F\left(p + \sum_{i=1}^{m} \sum_{j=1}^{\infty} \alpha_{ij}(u_{ij} - p)\right) - q = \sum_{i=1}^{m} \sum_{j=1}^{\infty} \alpha_{ij}\big(f(u_{ij}) - q\big).$$

Since $u_{ij} \in E_1$ for all i and j, it holds that $f(u_{ij}) \in f(E_1) = E_2$ for each i and j. Moreover, since $u_{ij} \in E_1 \cap B_r(p)$ for all i and j, it follows from Lemma 4.10 that

$$\begin{aligned}
\|f(u_{ij}) - q\|_a^2 &= \|(T_{-q} \circ f \circ T_p)(u_{ij} - p)\|_a^2 \\
&= \big\langle (T_{-q} \circ f \circ T_p)(u_{ij} - p), (T_{-q} \circ f \circ T_p)(u_{ij} - p) \big\rangle_a \\
&= \langle u_{ij} - p, u_{ij} - p \rangle_a \\
&= \|u_{ij} - p\|_a^2 \\
&< r^2
\end{aligned}$$

for all i and j. Hence, $f(u_{ij}) \in E_2 \cap B_r(q)$ for all i and j.

Furthermore, it follows from Lemma 4.10 that

$$\begin{aligned}
&\left\| \sum_{i=1}^{m} \sum_{j=1}^{\infty} \alpha_{ij} \big(f(u_{ij}) - q \big) \right\|_a^2 \\
&= \left\| \sum_{i=1}^{m} \sum_{j=1}^{\infty} \alpha_{ij} (T_{-q} \circ f \circ T_p)(u_{ij} - p) \right\|_a^2 \\
&= \left\langle \sum_{i=1}^{m} \sum_{j=1}^{\infty} \alpha_{ij} (T_{-q} \circ f \circ T_p)(u_{ij} - p), \sum_{k=1}^{m} \sum_{\ell=1}^{\infty} \alpha_{k\ell} (T_{-q} \circ f \circ T_p)(u_{k\ell} - p) \right\rangle_a \\
&= \sum_{i=1}^{m} \sum_{k=1}^{m} \sum_{j=1}^{\infty} \alpha_{ij} \sum_{\ell=1}^{\infty} \alpha_{k\ell} \big\langle (T_{-q} \circ f \circ T_p)(u_{ij} - p), (T_{-q} \circ f \circ T_p)(u_{k\ell} - p) \big\rangle_a \\
&= \sum_{i=1}^{m} \sum_{k=1}^{m} \sum_{j=1}^{\infty} \alpha_{ij} \sum_{\ell=1}^{\infty} \alpha_{k\ell} \langle u_{ij} - p, u_{k\ell} - p \rangle_a \\
&= \left\langle \sum_{i=1}^{m} \sum_{j=1}^{\infty} \alpha_{ij} (u_{ij} - p), \sum_{k=1}^{m} \sum_{\ell=1}^{\infty} \alpha_{k\ell} (u_{k\ell} - p) \right\rangle_a \\
&= \left\| \sum_{i=1}^{m} \sum_{j=1}^{\infty} \alpha_{ij} (u_{ij} - p) \right\|_a^2 \\
&< \infty,
\end{aligned}$$

since $\sum_{i=1}^{m} \sum_{j=1}^{\infty} \alpha_{ij}(u_{ij} - p) = x - p \in M_a$. Thus, on account of Remark 4.2, we see that $\sum_{i=1}^{m} \sum_{j=1}^{\infty} \alpha_{ij}(f(u_{ij}) - q) \in M_a$. Therefore, in view of Definition 4.7, we get

$$\begin{aligned}
y &= F(x) \\
&= q + \sum_{i=1}^{m} \sum_{j=1}^{\infty} \alpha_{ij} \big(f(u_{ij}) - q \big) \\
&\in \mathrm{GS}(E_2, q)
\end{aligned}$$

and we conclude that $F(\mathrm{GS}(E_1, p)) \subset \mathrm{GS}(E_2, q)$.

(c) Now we assume that $y \in \mathrm{GS}(E_2, q)$. By Definition 4.7, there exist some $m \in \mathbb{N}$, $v_{ij} \in E_2 \cap B_r(q)$, and some $\alpha_{ij} \in \mathbb{R}$ which satisfy the condition $y - q = \sum_{i=1}^{m} \sum_{j=1}^{\infty} \alpha_{ij}(v_{ij} - q) \in M_a$. Since $f : E_1 \to E_2$ is surjective, there exists a $u_{ij} \in E_1$

satisfying $v_{ij} = f(u_{ij})$ for any i and j. Moreover, by Lemma 4.10, we have

$$\begin{aligned}
\|u_{ij} - p\|_a^2 &= \langle u_{ij} - p, u_{ij} - p\rangle_a \\
&= \big\langle (T_{-q} \circ f \circ T_p)(u_{ij} - p), (T_{-q} \circ f \circ T_p)(u_{ij} - p)\big\rangle_a \\
&= \big\langle f(u_{ij}) - q, f(u_{ij}) - q\big\rangle_a \\
&= \langle v_{ij} - q, v_{ij} - q\rangle_a \\
&= \|v_{ij} - q\|_a^2 \\
&< r^2
\end{aligned}$$

for any i and j. So we conclude that $u_{ij} \in E_1 \cap B_r(p)$ and $v_{ij} = f(u_{ij})$ for all i and j.

On the other hand, using Lemma 4.10, we have

$$\begin{aligned}
&\left\| \sum_{i=1}^{m} \sum_{j=1}^{\infty} \alpha_{ij}(u_{ij} - p) \right\|_a^2 \\
&\quad = \left\langle \sum_{i=1}^{m} \sum_{j=1}^{\infty} \alpha_{ij}(u_{ij} - p), \sum_{k=1}^{m} \sum_{\ell=1}^{\infty} \alpha_{k\ell}(u_{k\ell} - p) \right\rangle_a \\
&\quad = \sum_{i=1}^{m} \sum_{k=1}^{m} \sum_{j=1}^{\infty} \alpha_{ij} \sum_{\ell=1}^{\infty} \alpha_{k\ell} \big\langle u_{ij} - p, u_{k\ell} - p \big\rangle_a \\
&\quad = \sum_{i=1}^{m} \sum_{k=1}^{m} \sum_{j=1}^{\infty} \alpha_{ij} \sum_{\ell=1}^{\infty} \alpha_{k\ell} \big\langle (T_{-q} \circ f \circ T_p)(u_{ij} - p), (T_{-q} \circ f \circ T_p)(u_{k\ell} - p)\big\rangle_a \\
&\quad = \left\langle \sum_{i=1}^{m} \sum_{j=1}^{\infty} \alpha_{ij}(T_{-q} \circ f \circ T_p)(u_{ij} - p), \sum_{k=1}^{m} \sum_{\ell=1}^{\infty} \alpha_{k\ell}(T_{-q} \circ f \circ T_p)(u_{k\ell} - p) \right\rangle_a \\
&\quad = \left\| \sum_{i=1}^{m} \sum_{j=1}^{\infty} \alpha_{ij}(T_{-q} \circ f \circ T_p)(u_{ij} - p) \right\|_a^2 \\
&\quad = \left\| \sum_{i=1}^{m} \sum_{j=1}^{\infty} \alpha_{ij}\big(f(u_{ij}) - q\big) \right\|_a^2 \\
&\quad = \left\| \sum_{i=1}^{m} \sum_{j=1}^{\infty} \alpha_{ij}(v_{ij} - q) \right\|_a^2 \\
&\quad < \infty,
\end{aligned}$$

since $\sum_{i=1}^{m}\sum_{j=1}^{\infty}\alpha_{ij}(v_{ij}-q)=y-q\in M_a$. Therefore, it follows from Remark 4.2 that $\sum_{i=1}^{m}\sum_{j=1}^{\infty}\alpha_{ij}(u_{ij}-p)\in M_a$.

Hence, it follows from Definition 4.11 that

$$\begin{aligned}
y &= q+\sum_{i=1}^{m}\sum_{j=1}^{\infty}\alpha_{ij}(f(u_{ij})-q)\\
&= q+\sum_{i=1}^{m}\sum_{j=1}^{\infty}\alpha_{ij}(T_{-q}\circ f\circ T_p)(u_{ij}-p)\\
&= q+(T_{-q}\circ F\circ T_p)\left(\sum_{i=1}^{m}\sum_{j=1}^{\infty}\alpha_{ij}(u_{ij}-p)\right)\\
&= F\left(p+\sum_{i=1}^{m}\sum_{j=1}^{\infty}\alpha_{ij}(u_{ij}-p)\right)\\
&\in F(\mathrm{GS}(E_1,p)).
\end{aligned}$$

Thus, we conclude that $\mathrm{GS}(E_2,q)\subset F(\mathrm{GS}(E_1,p))$.

(d) Similarly, referring to the changes presented in the tables below and following the previous parts (b) and (c) in this proof, we can prove that $\mathrm{GS}^2(E_2,q)=F_2(\mathrm{GS}^2(E_1,p))$.

The case $n=1$:	E_1	E_2	$\mathrm{GS}(E_1,p)$	$\mathrm{GS}(E_2,q)$	f
The case $n=2$:	$\mathrm{GS}(E_1,p)$	$\mathrm{GS}(E_2,q)$	$\mathrm{GS}^2(E_1,p)$	$\mathrm{GS}^2(E_2,q)$	F

The case $n=1$:	F	Definition 4.7	Definition 4.11	Lemma 4.10
The case $n=2$:	F_2	Definition 4.14 (i)	Definition 4.14 (ii)	(4.10)

(e) Finally, according to Proposition 4.18, Theorem 4.28, and (d), we further have

$$\mathrm{GS}^n(E_2,q)=\mathrm{GS}^2(E_2,q)=F_2(\mathrm{GS}^2(E_1,p))=F_n(\mathrm{GS}^n(E_1,p))$$

for any integer $n\geq 3$. □

4.7 Extension of Isometries to the Entire Space

Let $I^\omega=\prod_{i=1}^{\infty}I$ be the *Hilbert cube*, where $I=[0,1]$ is the unit closed interval. From now on, we assume that E_1 and E_2 are nonempty subsets of I^ω. It is clear that they are bounded.

In Theorem 4.34, we will prove that the domain E_1 of a local d_a-isometry $f : E_1 \to E_2$ can be extended to any real Hilbert space including the domain E_1 of f.

Definition 4.33. Let E_1 be a nonempty subset of I^ω that is d_a-isometric to a subset E_2 of I^ω via a surjective d_a-isometry $f : E_1 \to E_2$. Let p be an element of E_1 and q an element of E_2 with $q = f(p)$. Assume that $\{\frac{1}{a_i}e_i\}_{i\in\Lambda_\alpha}$ is a complete orthonormal sequence in the Hilbert space $\mathrm{GS}^2(E_1,p)-p$, where Λ_α is a nonempty proper subset of $\mathbb{N}$. Moreover, assume that $\{\beta_i\}_{i\in\mathbb{N}}$ is a complete orthonormal sequence in the Hilbert space M_a such that $\beta_i = \frac{1}{a_i}(T_{-q}\circ F_2\circ T_p)(e_i)$ for each $i \in \Lambda_\alpha$, where $F_2 : \mathrm{GS}^2(E_1,p) \to M_a$ is defined in Definition 4.14. Let p_i be the ith component of p, i.e., $p = \sum_{i=1}^{\infty} p_i e_i$. For any set Λ satisfying $\Lambda_\alpha \subset \Lambda \subset \mathbb{N}$, we define a basic cylinder or a basic interval $\tilde{J}$ by

$$\tilde{J} = \prod_{i=1}^{\infty} \tilde{J}_i, \quad \text{where} \quad \tilde{J}_i = \begin{cases} [0,1] & (\text{for } i \in \Lambda), \\ \{p_i\} & (\text{for } i \notin \Lambda). \end{cases}$$

Moreover, referring to Theorem 4.31, we define the function $G_2 : \mathrm{GS}^2(\tilde{J},p) \to M_a$ by

$$(T_{-q}\circ G_2\circ T_p)(u-p) = \sum_{i\in\Lambda(\tilde{J})} \left\langle u-p, \frac{1}{a_i}e_i \right\rangle_a \beta_i \tag{4.27}$$

for all $u \in \mathrm{GS}^2(\tilde{J},p)$.

The following theorem states that the domain of a local d_a-isometry can be extended to any real Hilbert space including the domain of the local d_a-isometry.

Theorem 4.34. *Let E_1 be a bounded subset of I^ω that contains at least two elements. Suppose E_1 is d_a-isometric to a subset E_2 of I^ω via a surjective d_a-isometry $f : E_1 \to E_2$. Let p and q be elements of E_1 and E_2 satisfying $q = f(p)$. Assume that $\{\frac{1}{a_i}e_i\}_{i\in\Lambda_\alpha}$ is a complete orthonormal sequence in the Hilbert space $\mathrm{GS}^2(E_1,p)-p$, where Λ_α is a nonempty proper subset of $\mathbb{N}$. Moreover, assume that $\{\beta_i\}_{i\in\mathbb{N}}$ is a complete orthonormal sequence in the Hilbert space M_a such that $\beta_i = \frac{1}{a_i}(T_{-q}\circ F_2\circ T_p)(e_i)$ for each $i \in \Lambda_\alpha$. Let Λ be a set satisfying $\Lambda_\alpha \subset \Lambda \subset \mathbb{N}$ and let $\tilde{J}$ be defined as in Definition 4.33. Then the function $G_2 : \mathrm{GS}^2(\tilde{J},p) \to M_a$ is a d_a-isometry and the function $T_{-q}\circ G_2\circ T_p : \mathrm{GS}^2(\tilde{J},p)-p \to M_a$ is linear. In particular, G_2 is an extension of F_2.*

Proof. (a) First, we assert that the function $T_{-q}\circ G_2\circ T_p : \mathrm{GS}^2(\tilde{J},p)-p \to M_a$ preserves the inner product. Assume that u and v are arbitrary elements of

$\mathrm{GS}^2(\tilde{J},p)$. Since $\Lambda = \Lambda(\tilde{J})$, it follows from (4.26), (4.27), and the orthonormality of $\{\frac{1}{a_i}e_i\}_{i\in\mathbb{N}}$ and $\{\beta_i\}_{i\in\mathbb{N}}$ that

$$\begin{aligned}
&\big\langle (T_{-q}\circ G_2\circ T_p)(u-p),\, (T_{-q}\circ G_2\circ T_p)(v-p)\big\rangle_a \\
&= \left\langle \sum_{i\in\Lambda}\left\langle u-p,\, \frac{1}{a_i}e_i\right\rangle_a \beta_i,\, \sum_{j\in\Lambda}\left\langle v-p,\, \frac{1}{a_j}e_j\right\rangle_a \beta_j\right\rangle_a \\
&= \sum_{i\in\Lambda}\left\langle u-p,\, \frac{1}{a_i}e_i\right\rangle_a \sum_{j\in\Lambda}\left\langle v-p,\, \frac{1}{a_j}e_j\right\rangle_a \langle \beta_i,\, \beta_j\rangle_a \\
&= \sum_{i\in\Lambda}\left\langle u-p,\, \frac{1}{a_i}e_i\right\rangle_a \sum_{j\in\Lambda}\left\langle v-p,\, \frac{1}{a_j}e_j\right\rangle_a \left\langle \frac{1}{a_i}e_i,\, \frac{1}{a_j}e_j\right\rangle_a \\
&= \left\langle \sum_{i\in\Lambda}\left\langle u-p,\, \frac{1}{a_i}e_i\right\rangle_a \frac{1}{a_i}e_i,\, \sum_{j\in\Lambda}\left\langle v-p,\, \frac{1}{a_j}e_j\right\rangle_a \frac{1}{a_j}e_j\right\rangle_a \\
&= \langle u-p, v-p\rangle_a
\end{aligned}$$

for all $u, v \in \mathrm{GS}^2(\tilde{J},p)$, i.e., $T_{-q}\circ G_2\circ T_p$ preserves the inner product.

(b) We assert that G_2 is a d_a-isometry. Let u and v be arbitrary elements of $\mathrm{GS}^2(\tilde{J},p)$. Since $T_{-q}\circ G_2\circ T_p$ preserves the inner product by (a), we have

$$\begin{aligned}
&d_a\big(G_2(u), G_2(v)\big)^2 \\
&= \big\|(T_{-q}\circ G_2\circ T_p)(u-p) - (T_{-q}\circ G_2\circ T_p)(v-p)\big\|_a^2 \\
&= \big\langle (T_{-q}\circ G_2\circ T_p)(u-p) - (T_{-q}\circ G_2\circ T_p)(v-p), \\
&\qquad (T_{-q}\circ G_2\circ T_p)(u-p) - (T_{-q}\circ G_2\circ T_p)(v-p)\big\rangle_a \\
&= \langle u-p, u-p\rangle_a - \langle u-p, v-p\rangle_a - \langle v-p, u-p\rangle_a + \langle v-p, v-p\rangle_a \\
&= \big\langle (u-p)-(v-p), (u-p)-(v-p)\big\rangle_a \\
&= \|(u-p)-(v-p)\|_a^2 \\
&= \|u-v\|_a^2 \\
&= d_a(u,v)^2
\end{aligned}$$

for all $u, v \in \mathrm{GS}^2(\tilde{J},p)$, i.e., $G_2 : \mathrm{GS}^2(\tilde{J},p) \to M_a$ is a d_a-isometry.

(c) We assert that the function $T_{-q}\circ G_2\circ T_p : \mathrm{GS}^2(\tilde{J},p) - p \to M_a$ is linear. Assume that u and v are arbitrary elements of $\mathrm{GS}^2(\tilde{J},p)$ and α, β are real numbers. Since $\mathrm{GS}^2(\tilde{J},p) - p$ is a real vector space, it holds that $\alpha(u-p)+\beta(v-p) \in \mathrm{GS}^2(\tilde{J},p) - p$. Thus, $\alpha(u-p)+\beta(v-p) = w-p$ for some $w \in \mathrm{GS}^2(\tilde{J},p)$. Hence, referring to the changes presented in the table below and following (d) of the proof of Theorem 4.13, we can easily prove that $T_{-q}\circ G_2\circ T_p$ is linear.

Theorem 4.13:	$\mathrm{GS}(E_1,p)$	F	(4.10)
Here:	$\mathrm{GS}^2(\tilde{J},p)$	G_2	(a)

(d) Finally, we assert that G_2 is an extension of F_2. Let $\hat{J}$ be either a basic cylinder or a basic interval defined by

$$\hat{J} = \prod_{i=1}^{\infty} \hat{J}_i, \quad \text{where} \quad \hat{J}_i = \begin{cases} [0,1] & (\text{for } i \in \Lambda_\alpha), \\ \{p_i\} & (\text{for } i \notin \Lambda_\alpha). \end{cases}$$

We see that $p = (p_1, p_2, \ldots) \in \hat{J} \cap E_1$ and $\Lambda(\hat{J}) = \Lambda_\alpha = \Lambda(\mathrm{GS}^2(E_1,p))$.

By Lemma 4.19, if $i \in \Lambda(\mathrm{GS}^2(E_1,p))$, then $\alpha_i e_i \in \mathrm{GS}^2(E_1,p) - p$ for all $\alpha_i \in \mathbb{R}$. Since $\mathrm{GS}^2(E_1,p) - p$ is a real vector space, if we set $\Lambda_n = \{i \in \Lambda(\mathrm{GS}^2(E_1,p)) : i < n\}$, then we have

$$\sum_{i \in \Lambda_n} \alpha_i e_i \in \mathrm{GS}^2(E_1,p) - p$$

for all $n \in \mathbb{N}$ and for all $\alpha_i \in \mathbb{R}$. For now, with all α_i's fixed, we define $x_n = p + \sum_{i \in \Lambda_n} \alpha_i e_i$ for any $n \in \mathbb{N}$. Then $\{x_n\}$ is a sequence in $\mathrm{GS}^2(E_1,p)$. When $\{x_n\}$ converges in M_a, it holds that

$$p + \sum_{i \in \Lambda(\mathrm{GS}^2(E_1,p))} \alpha_i e_i = \lim_{n \to \infty} x_n \in \mathrm{GS}^2(E_1,p),$$

because $\mathrm{GS}^2(E_1,p)$ is closed by Lemma 4.17 (iii). That is,

$$\left\{ p + \sum_{i \in \Lambda(\mathrm{GS}^2(E_1,p))} \alpha_i e_i \in M_a : \alpha_i \in \mathbb{R} \text{ for all } i \in \Lambda\big(\mathrm{GS}^2(E_1,p)\big) \right\} \subset \mathrm{GS}^2(E_1,p).$$

Hence, by the previous inclusion and Theorem 4.24 with $\beta = \left\{\frac{1}{a_i} e_i\right\}_{i \in \mathbb{N}}$ and $J_\beta = \hat{J}$, we get

$$\begin{aligned} \mathrm{GS}(\hat{J},p) - p &= \left\{ \sum_{i \in \Lambda(\hat{J})} \alpha_i e_i \in M_a : \alpha_i \in \mathbb{R} \text{ for all } i \in \Lambda(\hat{J}) \right\} \\ &= \left\{ \sum_{i \in \Lambda(\mathrm{GS}^2(E_1,p))} \alpha_i e_i \in M_a : \alpha_i \in \mathbb{R} \text{ for all } i \in \Lambda\big(\mathrm{GS}^2(E_1,p)\big) \right\} \\ &\subset \mathrm{GS}^2(E_1,p) - p. \end{aligned}$$

So we have

$$\hat{J} \cap B_r(p) \subset \mathrm{GS}(\hat{J}, p) \cap B_r(p) \subset \mathrm{GS}^2(E_1, p) \cap B_r(p)$$

for some real number $r > 0$ and hence, we further have

$$\mathrm{GS}(\hat{J}, p) \subset \mathrm{GS}^2(\hat{J}, p) \subset \mathrm{GS}^3(E_1, p) = \mathrm{GS}^2(E_1, p).$$

Moreover, by Remark 4.30, we know that $\mathrm{GS}^2(\hat{J}, p) = \mathrm{GS}(\hat{J}, p)$. Hence, we have

$$\mathrm{GS}(\hat{J}, p) = \mathrm{GS}^2(\hat{J}, p) \subset \mathrm{GS}^2(E_1, p).$$

On the other hand, since $\left\{\frac{1}{a_i} e_i\right\}_{i \in \Lambda_\alpha}$ is a complete orthonormal sequence in $\mathrm{GS}^2(E_1, p) - p$, it follows from Theorem 4.24 with $\beta = \left\{\frac{1}{a_i} e_i\right\}_{i \in \mathbb{N}}$ that

$$x = \sum_{i \in \Lambda_\alpha} \left\langle x, \frac{1}{a_i} e_i \right\rangle_a \frac{1}{a_i} e_i \in \mathrm{GS}(\hat{J}, p) - p = \mathrm{GS}^2(\hat{J}, p) - p$$

for all $x \in \mathrm{GS}^2(E_1, p) - p$, which implies that $\mathrm{GS}^2(E_1, p) = \mathrm{GS}^2(\hat{J}, p) = \mathrm{GS}(\hat{J}, p)$.

Let u be an arbitrary element of $\mathrm{GS}^2(E_1, p)$. Then by (4.26) with $\hat{J}$ instead of J, we have

$$u - p = \sum_{i \in \Lambda(\hat{J})} \left\langle u - p, \frac{1}{a_i} e_i \right\rangle_a \frac{1}{a_i} e_i \tag{4.28}$$

and since $T_{-q} \circ F_2 \circ T_p$ is linear and continuous, we use (4.27), (4.28), and the facts $\mathrm{GS}^2(E_1, p) = \mathrm{GS}^2(\hat{J}, p) = \mathrm{GS}(\hat{J}, p)$ and $\Lambda(\hat{J}) = \Lambda_\alpha = \Lambda(\mathrm{GS}^2(E_1, p))$ to have

$$\begin{aligned}
&(T_{-q} \circ G_2 \circ T_p)(u - p) \\
&\quad = \sum_{i \in \Lambda(\tilde{J})} \left\langle u - p, \frac{1}{a_i} e_i \right\rangle_a \beta_i \\
&\quad = \sum_{i \in \Lambda(\hat{J})} \left\langle u - p, \frac{1}{a_i} e_i \right\rangle_a \beta_i \\
&\quad = \sum_{i \in \Lambda(\hat{J})} \left\langle u - p, \frac{1}{a_i} e_i \right\rangle_a \frac{1}{a_i} (T_{-q} \circ F_2 \circ T_p)(e_i) \\
&\quad = \lim_{n \to \infty} \sum_{i \in \Lambda_n(\hat{J})} \left\langle u - p, \frac{1}{a_i} e_i \right\rangle_a \frac{1}{a_i} (T_{-q} \circ F_2 \circ T_p)(e_i) \\
&\quad = \lim_{n \to \infty} (T_{-q} \circ F_2 \circ T_p) \left(\sum_{i \in \Lambda_n(\hat{J})} \left\langle u - p, \frac{1}{a_i} e_i \right\rangle_a \frac{1}{a_i} e_i \right) \\
&\quad = (T_{-q} \circ F_2 \circ T_p) \left(\sum_{i \in \Lambda(\hat{J})} \left\langle u - p, \frac{1}{a_i} e_i \right\rangle_a \frac{1}{a_i} e_i \right) \\
&\quad = (T_{-q} \circ F_2 \circ T_p)(u - p),
\end{aligned}$$

where we set $\Lambda_n(\hat{J}) = \{i \in \Lambda(\hat{J}) : i < n\}$ for every $n \in \mathbb{N}$.

Therefore, it follows that $(T_{-q} \circ G_2 \circ T_p)(u - p) = (T_{-q} \circ F_2 \circ T_p)(u - p)$ for all $u \in \mathrm{GS}^2(E_1, p)$, i.e., $G_2(u) = F_2(u)$ for all $u \in \mathrm{GS}^2(E_1, p)$. In other words, G_2 is an extension of F_2. □

In Theorem 4.34, $\mathrm{GS}^2(E_1, p) - p$ is closed in M_a. Hence, $\mathrm{GS}^2(E_1, p) - p$ is a closed subspace of the real Hilbert space M_a, i.e., $\mathrm{GS}^2(E_1, p) - p$ is itself a real Hilbert space. Similarly, $\mathrm{GS}^2(\tilde{J}, p) - p$ is a real Hilbert space. Since the d_a-isometry $T_{-q} \circ G_2 \circ T_p : \mathrm{GS}^2(\tilde{J}, p) - p \to G_2(\mathrm{GS}^2(\tilde{J}, p)) - q$ is a homeomorphism and $(T_{-q} \circ G_2 \circ T_p)(\mathrm{GS}^2(\tilde{J}, p) - p) = G_2(\mathrm{GS}^2(\tilde{J}, p)) - q$, $G_2(\mathrm{GS}^2(\tilde{J}, p)) - q$ is a closed subspace of the real Hilbert space M_a, i.e., it is also a real Hilbert space. Similarly, $\mathrm{GS}^2(E_2, q) - q$ is a real Hilbert space.

Assuming the axiom of choice, it is well known that

(i) Every Hilbert space has a complete orthonormal sequence (ref. [2, p. 140]).

(ii) The union of a complete orthonormal sequence in the Hilbert space H and a complete orthonormal sequence in the orthogonal complement of H is a complete orthonormal sequence in the Hilbert space $H \oplus H^{\perp}$ (see Remark 2.46 (ii) or [3, p. 124]).

Indeed, $\{\beta_i\}_{i\in\Lambda_\alpha} = \left\{\frac{1}{a_i}(T_{-q} \circ F_2 \circ T_p)(e_i)\right\}_{i\in\Lambda_\alpha}$ is a complete orthonormal sequence in the real Hilbert space $\mathrm{GS}^2(E_2, q) - q$. In view of (i), we can choose a complete orthonormal sequence $\{\beta_i\}_{i\in\Lambda\setminus\Lambda_\alpha}$ in the real Hilbert space $(\mathrm{GS}^2(E_2, q) - q)^{\perp} \cap (G_2(\mathrm{GS}^2(\tilde{J}, p)) - q)$. We can also choose a complete orthonormal sequence $\{\beta_i\}_{i\in\mathbb{N}\setminus\Lambda}$ in the orthogonal complement of $G_2(\mathrm{GS}^2(\tilde{J}, p)) - q$.

We note that

$$\begin{aligned} M_a = (\mathrm{GS}^2(E_2, q) - q) &\oplus \big((\mathrm{GS}^2(E_2, q) - q)^{\perp} \cap (G_2(\mathrm{GS}^2(\tilde{J}, p)) - q)\big) \\ &\oplus (G_2(\mathrm{GS}^2(\tilde{J}, p)) - q)^{\perp}, \end{aligned}$$

where $(\mathrm{GS}^2(E_2, q) - q)^{\perp}$ and $(G_2(\mathrm{GS}^2(\tilde{J}, p)) - q)^{\perp}$ are the orthogonal complements of the Hilbert spaces $\mathrm{GS}^2(E_2, q) - q$ and $G_2(\mathrm{GS}^2(\tilde{J}, p)) - q$, respectively. Since $\mathrm{GS}^2(E_1, p) \subset \mathrm{GS}^2(\tilde{J}, p)$, it follows from Theorem 4.32 that $\mathrm{GS}^2(E_2, q) = F_2(\mathrm{GS}^2(E_1, p)) \subset G_2(\mathrm{GS}^2(\tilde{J}, p)) \subset M_a$. Hence,

$$\{\beta_i\}_{i\in\mathbb{N}} = \left\{\frac{1}{a_i}(T_{-q} \circ F_2 \circ T_p)(e_i)\right\}_{i\in\Lambda_\alpha} \cup \{\beta_i\}_{i\in\Lambda\setminus\Lambda_\alpha} \cup \{\beta_i\}_{i\in\mathbb{N}\setminus\Lambda}$$

is a complete orthonormal sequence in the real Hilbert space M_a.

The pair (H_1, H_2) of Hilbert spaces is said to have the *isometric extension property* if for every isometry f from an arbitrary subset S of H_1 into H_2, there exists an isometry F of H_1 into H_2 such that the restriction of F to S coincides with f.

The following theorem is a well known result due to [23, Theorem 11.4].

Theorem 4.35 (Wells and Williams). *If H is a Hilbert space, then (H, H) has the isometric extension property if and only if H is finite-dimensional. In general, if $S \subset H$ and $f : S \to H$ is an isometry, then f can be extended as an isometry to the closed linear span of S.*

We note that Theorem 4.35 does not imply Theorem 4.34. For example, assume that E_1 and E_2 are subsets of the Hilbert cube I^ω and Λ is a proper subset of $\mathbb{N}$ and a proper superset of Λ_α. Then $\mathrm{GS}^2(\tilde{J}, p) - p$ is a proper subspace of the real Hilbert space M_a and $\mathrm{GS}^2(E_1, p) - p$ is a proper subspace of $\mathrm{GS}^2(\tilde{J}, p) - p$. Nevertheless, it follows from Theorem 4.34 that every surjective isometry $f : E_1 \to E_2$ can be extended to an isometry $G_2 : \mathrm{GS}^2(\tilde{J}, p) \to M_a$. On the other hand, we cannot expect to obtain this result using Theorem 4.35, since the closed linear span of E_1 is a proper subset of $\mathrm{GS}^2(\tilde{J}, p) - p$, which implies that Theorem 4.34 is not only different from Theorem 4.35 but also has many advantages over it.

Moreover, for any bounded subset S of I^ω, it is clear that $\overline{\mathrm{span}}(S) \subset \mathrm{GS}^2(S, p)$. But it is not yet clear whether $\overline{\mathrm{span}}(S) = \mathrm{GS}^2(S, p)$, where $\overline{\mathrm{span}}(S)$ denotes the closed linear span of S. If $\overline{\mathrm{span}}(S) \neq \mathrm{GS}^2(S, p)$ is correct, Theorem 4.34 has more advantages than Theorem 4.35.

According to Theorem 4.34, the domain of a local d_a-isometry can be extended to any real Hilbert space containing that domain.

Chapter 5

History of Ulam's Conjecture

A conjecture of Ulam states that the standard product probability measure π on the Hilbert cube I^ω is d_a-invariant when the sequence $a = \{a_i\}_{i \in \mathbb{N}}$ of positive numbers satisfies the condition $\sum_{i=1}^{\infty} a_i^2 < \infty$. In 1974 and 1977, J. Mycielski published the first papers on this topic. Indeed, he proved the conjecture of Ulam affirmatively under the additional assumption that the sets are open. In 1982, J. W. Fickett succeeded in partially proving Ulam's conjecture by proving the following statement in a different way than J. Mycielski: If the a_i's tend to 0 very rapidly, then any two Borel subsets of the Hilbert cube which are d_a-isometric have the same standard product probability measure. About 40 years later, in 2018, S.-M. Jung and E. Kim jointly studied Ulam's conjecture and improved the result of J. W. Fickett by partially proving Ulam's conjecture. In this chapter, we will introduce the historical process of solving Ulam's conjecture by presenting a summary of these papers. In some cases, the reader may skip reading this chapter.

5.1 Historical Background

S. M. Ulam raised the conjecture on the invariance of measures defined in the compact metric space (see [21]):

> *Let X be a compact metric space. Does there exist a finitely additive measure μ defined for at least all the Borel subsets of X, such that $\mu(X) = 1$, $\mu(p) = 0$ for all points p of X, and such that congruent subsets of X have equal measure?*

Thereafter, J. Mycielski [17] confined the question of Ulam to the Hilbert cube I^ω and reformulated it using modern mathematical terms:

S.-M. Jung, *Ulam's Conjecture on Invariance of Measure in the Hilbert Cube*, Frontiers in Mathematics, https://doi.org/10.1007/978-3-031-30886-4_5

The standard product probability measure π on I^ω is d_a-invariant.

The above statement is widely known today as the *conjecture of Ulam*. (For the definition of the standard product probability measure, see Definition 3.38.)

In 1974, by using the axiom of choice, J. Mycielski [16, 17] answered the question of Ulam affirmatively under the additional assumption that the sets are open. In addition, he asked in [16] whether one can prove the conjecture of Ulam under the assumption that the sets are closed. J. W. Fickett [4], one step further, showed in 1982 that Ulam's conjecture is true when the sequence $a = \{a_i\}_{i \in \mathbb{N}}$ decreases very rapidly to 0 such that

$$\frac{a_{i+1}^{1/2^{i+1}}}{a_i} \to 0 \quad \text{as} \quad i \to \infty.$$

In 2018, S.-M. Jung and E. Kim proved in their paper [10] that Ulam's conjecture is true when the sequence $a = \{a_i\}_{i \in \mathbb{N}}$ of positive real numbers is monotone decreasing and satisfies the condition

$$a_{i+1} = o\left(\frac{a_i}{\sqrt{i}}\right) \quad \text{as} \quad i \to \infty$$

(see also [5, 11]). It is evident that the last condition is much weaker than that of Fickett.

In 2020, S.-M. Jung combined various methods to prove Ulam's conjecture, among which he developed and applied a method to extend the domain of local isometries, and he was able to completely prove that Ulam's conjecture is true. A detailed explanation of this is deferred to the last chapter.

5.2 Basic Definitions

We denote by $\mathbb{R}^\omega$ the infinite-dimensional real vector space defined as

$$\mathbb{R}^\omega = \big\{(x_1, x_2, \ldots) : x_i \in \mathbb{R} \text{ for all } i \in \mathbb{N}\big\}.$$

In addition, $(\mathbb{R}^\omega, \mathcal{T})$ denotes the product space $\prod_{i=1}^{\infty} \mathbb{R}$, where $(\mathbb{R}, \mathcal{T}_{\mathbb{R}})$ is the usual topological space.

Let $I = [0, 1]$ be the unit closed interval and $I^\omega = \prod_{i=1}^{\infty} I$ the *Hilbert cube*, and let π be the *standard product probability measure* on I^ω. We denote by $(I^\omega, \mathcal{T}_\omega)$ the (topological) subspace of $(\mathbb{R}^\omega, \mathcal{T})$. Then, $\mathcal{T}_\omega$ is the relative topology for I^ω induced by $\mathcal{T}$.

As in Sect. 4.1, let $a = \{a_i\}_{i\in\mathbb{N}}$ be a sequence of positive real numbers that satisfies the condition (4.1). Using this sequence a, we define the metric on I^ω by formula (4.2), i.e.,

$$d_a(x, y) = \left(\sum_{i=1}^{\infty} a_i^2 (x_i - y_i)^2\right)^{1/2}$$

for all $x = (x_1, x_2, \ldots) \in I^\omega$ and $y = (y_1, y_2, \ldots) \in I^\omega$.

As in Sect. 4.1 (see Theorem 4.3), we define the real Hilbert space M_a by

$$M_a = \left\{(x_1, x_2, \ldots) \in \mathbb{R}^\omega : \sum_{i=1}^{\infty} a_i^2 x_i^2 < \infty\right\}$$

and we define an inner product $\langle\cdot,\cdot\rangle_a$ on M_a by

$$\langle x, y\rangle_a = \sum_{i=1}^{\infty} a_i^2 x_i y_i$$

for all $x = (x_1, x_2, \ldots)$ and $y = (y_1, y_2, \ldots)$ of M_a. Then this inner product induces the norm

$$\|x\|_a = \sqrt{\langle x, x\rangle_a}$$

for all $x \in M_a$.

In view of definition (4.2), the metric d_a on I^ω can be extended to the metric on M_a, i.e.,

$$d_a(x, y) = \sqrt{\langle x - y, x - y\rangle_a}$$

for all $x, y \in M_a$.

5.3 Mycielski's Partial Solution

Using the axiom of choice, J. Mycielski [16, 17] answered Ulam's question in the affirmative with the additional assumption that the sets are open.

Definition 5.1. Let (X, d) be a metric space. We define an *entropy* by

$$E(K, t) = \min\left\{\operatorname{card}\mathcal{U} : \mathcal{U} \text{ is a } t\text{-covering of } K\right\}$$

for all compact subsets K of X and $0 < t \leq 1$.

(i) A compact subset K of X is called *thin* if for any $\varepsilon > 0$ there exist a $\delta > 0$ and an open subset U of X including K such that

$$E(C,t) \leq \varepsilon E(X,t)$$

for all compact subsets C of X with $C \subset U$ and for all $0 < t < \delta$.

(ii) A nonempty compact subset K of X is called *thick* in X if there exists an open subset U of X including K and a real number $\alpha > 0$ such that

$$E(C,t) \leq \alpha E(K,t)$$

for all compact subsets C of X with $C \subset U$ and for all $0 < t \leq 1$.

Now we introduce the definitions of filter and ultrafilter.

Definition 5.2. Let X be a set. A nonempty collection $\mathcal{F}$ of nonempty subsets of X is called a *filter* on X if it has the properties:

(i) If $A, B \in \mathcal{F}$, then $A \cap B \in \mathcal{F}$.

(ii) If $A \in \mathcal{F}$ and $A \subset B$, then $B \in \mathcal{F}$.

A filter $\mathcal{F}$ on X is called a *proper filter* if it is different from the power set $\mathcal{P}(X)$ of X. A proper filter $\mathcal{F}$ on X is said to be an *ultrafilter* or a *maximal filter* if no other proper filter on X contains $\mathcal{F}$ as a proper subset. An ultrafilter $\mathcal{F}$ is called *principal* if

$$\ker\mathcal{F} = \bigcap_{A \in \mathcal{F}} A = \{x\} \in \mathcal{F}$$

for some $x \in X$.

Let X be a topological space. For any element $x \in X$, the *neighborhood system* $\mathcal{N}_x$ of x is called the *neighborhood filter* of x.

Definition 5.3. Let $(X, \mathcal{T})$ be a topological space, let $x \in X$, and let $\mathcal{F}$ be a filter on X. Then $\mathcal{F}$ is said to converge to x if $\mathcal{F}$ contains the neighborhood filter of x.

For any element x of a topological space X, the neighborhood filter $\mathcal{N}_x$ of x converges to x.

Remark 5.4. Let $[0, \infty]$ be the topological space equipped with the natural compact topology and let $\mathcal{F}$ be an ultrafilter of subsets of $\mathbb{N}$. For each sequence $\{x_i\}_{i \in \mathbb{N}}$, let $\lim_{i \to \mathcal{F}} x_i$ be the unique $x \in [0, \infty]$ such that $\{i \in \mathbb{N} : x_i \in U\} \in \mathcal{F}$ for any neighborhood U of x. This generalized limit has the following properties:

(i) For each given $n \in \mathbb{N}$, let $f : [0,\infty]^n \to [0,\infty]$ be a continuous function. If $x_{i,j} \in [0,\infty]$ for all $i \in \mathbb{N}$ and $j \in \{1,2,\ldots,n\}$, then

$$\lim_{i\to\mathcal{F}} f(x_{i,1}, x_{i,2}, \ldots, x_{i,n}) = f\Big(\lim_{i\to\mathcal{F}} x_{i,1}, \lim_{i\to\mathcal{F}} x_{i,2}, \ldots, \lim_{i\to\mathcal{F}} x_{i,n}\Big).$$

(ii) For any sequences $\{x_i\}_{i\in\mathbb{N}}$ and $\{y_i\}_{i\in\mathbb{N}}$ in $[0,\infty]$, it holds that

$$\lim_{i\to\mathcal{F}} (x_i + y_i) = \lim_{i\to\mathcal{F}} x_i + \lim_{i\to\mathcal{F}} y_i,$$

where we set $x + \infty = \infty$ for all $x \in [0,\infty]$.

We remember that the *Borel measure* is the measure μ defined on the σ-algebra of Borel sets.

Theorem 5.5. *Let (X,d) be a metric space. If K is a compact subset of X that is thick in X, then there exists a Borel measure μ on X with the properties:*

(i) $\mu(K) = 1$.

(ii) *If U and V are open subsets of X that are isometric to each other, then $\mu(U) = \mu(V)$.*

Proof. Let $\mathcal{C}$ be the collection of all compact subsets of X. Then it holds that $K \in \mathcal{C}$.

(a) We assert that there exists a function $\lambda : \mathcal{C} \to [0,\infty]$ with the following properties:

(i) $\lambda(K) = 1$.

(ii) There are an open subset U of X and a constant $0 < \alpha < \infty$ such that $K \subset U$ and $\lambda(A) < \alpha$ for all $A \in \mathcal{C}$ with $A \subset U$.

(iii) $\lambda(A \cup B) \le \lambda(A) + \lambda(B)$ for all $A, B \in \mathcal{C}$.

(iv) $\lambda(A \cup B) = \lambda(A) + \lambda(B)$ for all $A, B \in \mathcal{C}$ with $A \cap B = \emptyset$.

(v) $\lambda(A) = \lambda(B)$ for all $A, B \in \mathcal{C}$ that are isometric to each other.

For the proof of the assertion above, we put

$$\lambda(A) = \lim_{i\to\mathcal{F}} \frac{E(A, 1/i)}{E(K, 1/i)}$$

for all $A \in \mathcal{C}$, where $\mathcal{F}$ is a non-principal ultrafilter of subsets of $\mathbb{N}$. All properties from (i) to (v) are obvious from this definition and the assumption that K is thick.

(b) We define

$$\mu_0(U) = \sup\{\lambda(A) : A \subset U \text{ and } A \in \mathcal{C}\},$$

for all open subsets U of X, and

$$\mu^*(V) = \inf\{\mu_0(U) : V \subset U \text{ and } U \text{ is open in } X\}$$

for any subset V of X.

We assert that μ^* has the following properties:

(i) $\mu^*(\emptyset) = 0$.

(ii) $1 \leq \mu^*(K) < \infty$.

(iii) If V_1 and V_2 are subsets of X with $V_1 \subset V_2$, then $\mu^*(V_1) \leq \mu^*(V_2)$.

(iv) If $\{V_i\}_{i\in\mathbb{N}}$ is a sequence of subsets of X, then

$$\mu^*\left(\bigcup_{i=1}^{\infty} V_i\right) \leq \sum_{i=1}^{\infty} \mu^*(V_i).$$

(v) If V_1 and V_2 are subsets of X that are positively separated, then $\mu^*(V_1 \cup V_2) = \mu^*(V_1) + \mu^*(V_2)$.

(vi) If U_1 and U_2 are open subsets of X that are isometric to each other, then $\mu^*(U_1) = \mu^*(U_2)$.

Due to (i) in (a), if we put $A = K$ and $B = \emptyset$ in (iv) of (a), then we have $\lambda(\emptyset) = 0$. Thus, (i) is true. Furthermore, (ii) is true by (i) and (ii) of (a). (iii) is obviously true by the definition of μ^*.

Now we focus on the proof of (iv). We prove that

$$\mu_0\left(\bigcup_{i=1}^{\infty} U_i\right) \leq \sum_{i=1}^{\infty} \mu_0(U_i) \tag{5.1}$$

for any sequence $\{U_i\}_{i\in\mathbb{N}}$ of open subsets of X.

Assume that $A \in \mathcal{C}$ with

$$A \subset \bigcup_{i=1}^{\infty} U_i.$$

Then, since A is a compact subset of X, there exists an $m \in \mathbb{N}$ such that

$$A \subset \bigcup_{i=1}^{m} U_i$$

and there exist sets $A_1, A_2, \ldots, A_m \in \mathcal{C}$ such that

$$A = \bigcup_{i=1}^{m} A_i \quad \text{and} \quad A_j \subset U_j$$

for each $j \in \{1, 2, \ldots, m\}$. Hence, the inequality (5.1) follows from (iii) of (a) and the definition of μ_0.

Now, for any $\varepsilon > 0$, we can choose open subsets U_i of X with $V_i \subset U_i$ and

$$\mu_0(U_i) \leq \mu^*(V_i) + \frac{1}{2^i}\varepsilon.$$

Then it follows from (5.1) that

$$\mu^*\left(\bigcup_{i=1}^{\infty} V_i\right) \leq \mu_0\left(\bigcup_{i=1}^{\infty} U_i\right) \leq \sum_{i=1}^{\infty} \mu_0(U_i) \leq \sum_{i=1}^{\infty} \mu^*(V_i) + \varepsilon,$$

which implies the validity of (iv).

To the proof of (v), since V_1 and V_2 are positively separated, there exist open subsets U_1 and U_2 of X such that $V_1 \subset U_1$, $V_2 \subset U_2$, and $U_1 \cap U_2 = \emptyset$. For each $A \in \mathcal{C}$ with $A \subset U_1 \cup U_2$, we have $A \cap U_1 \in \mathcal{C}$ and $A \cap U_2 \in \mathcal{C}$. Thus, by (iv) in (a), we get $\mu^*(V_1 \cup V_2) \geq \mu^*(V_1) + \mu^*(V_2)$ and (v) follows from (iv).

Finally, (vi) follows from (v) of (a).

(c) Due to (i), (iii), (iv), and (v) of (b), μ^* is a metric outer measure on X. Thus, by Theorem 3.30, every Borel set in X is μ^*-measurable. In view of (ii) in (b), we can write

$$\mu(A) = \frac{\mu^*(A)}{\mu^*(K)}$$

for all $A \in \mathcal{B}$, where we use $\mathcal{B}$ to denote the σ-algebra of all Borel sets in X.

Therefore, μ is a Borel measure on $\mathcal{B}$ and $\mu(K) = 1$. The last requirement of our theorem is satisfied by (vi) of (b). □

According to Remark 4.1 (iii), $(I^\omega, \mathcal{T}_\omega)$ is a compact subspace of $(\mathbb{R}^\omega, \mathcal{T})$. Thus, I^ω is a compact subspace of I^ω. Moreover, we note that $\mathcal{T}_\omega$ is the topology for I^ω generated by the metric d_a and I^ω is thick in itself. If we substitute I^ω for both X and K in Theorem 5.5, then we immediately obtain (i) and (iii) of the following theorem.

Theorem 5.6. *There exists a Borel measure μ on I^ω with the following properties:*

(i) $\mu(I^\omega) = 1$.

(*ii*) *If K is a thin subset of I^ω, then $\mu(K) = 0$.*

(*iii*) *If U and V are open subsets of I^ω that are d_a-isometric to each other, then $\mu(U) = \mu(V)$.*

In particular, the measure μ is the Borel measure given in Theorem 5.5.

Proof. We only have to prove (ii). Assume that K is a thin subset of I^ω. In view of Definition 5.1 (i), for any $\varepsilon > 0$, there exist a $\delta_\varepsilon > 0$ and an open subset U_ε of I^ω such that $K \subset U_\varepsilon$ and

$$E(C,t) \leq \varepsilon E(I^\omega, t)$$

for all compact subsets C of U_ε and $0 < t < \delta_\varepsilon$.

Furthermore, we get

$$\lambda(C) = \lim_{i \to \mathcal{F}} \frac{E(C, 1/i)}{E(I^\omega, 1/i)} \leq \varepsilon$$

for all compact subsets C of U_ε, where λ is defined in (a) of the proof of Theorem 5.5, which implies that $\lambda(C) = 0$ for all compact subsets C of U_ε. Therefore, it holds that

$$\mu_0(U_\varepsilon) = \sup\left\{\lambda(C) : C \text{ is a compact subset of } U_\varepsilon\right\} = 0,$$

where μ_0 is defined in (b) of the proof of Theorem 5.5.

Finally, by (c) in the proof of Theorem 5.5, we obtain

$$\begin{aligned}
\mu(K) &= \frac{\mu^*(K)}{\mu^*(I^\omega)} \\
&= \frac{\inf\left\{\mu_0(U) : K \subset U \text{ and } U \text{ is open in } I^\omega\right\}}{\inf\left\{\mu_0(U) : I^\omega \subset U \text{ and } U \text{ is open in } I^\omega\right\}} \\
&\leq \frac{\mu_0(U_\varepsilon)}{\mu_0(I^\omega)} \\
&= 0,
\end{aligned}$$

as required. □

Proposition 5.7. *If K is a compact subset of I^ω and for each $n \in \mathbb{N}$ there exists an open subset U of I^ω such that $K \subset U$ and there are disjoint subsets $U_1, U_2, \ldots, U_n$ of I^ω, all d_a-isometric to U, then K is thin.*

Proof. Given any $\varepsilon > 0$, we select an integer n not less than $\frac{1}{\varepsilon}$. Let U be an open subset of I^ω such that $K \subset U$ and there exist disjoint subsets $U_1, U_2, \ldots, U_n$ of I^ω, all d_a-isometric to U. Then for any compact subset C of U, there are n disjoint compact subsets C_i of U_i, all d_a-isometric to C. We set

$$\delta = \frac{1}{2} \min \{d_a(x, y) : x \in C_i \text{ and } y \in C_j \text{ for } i \neq j\}.$$

Then, we have

$$E(C, t) \leq \frac{1}{n} E(I^\omega, t) \leq \varepsilon E(I^\omega, t)$$

for all $0 < t < \delta$. Thus, it follows that K is thin. □

We note that the metric d_a on the Hilbert cube I^ω is translation-invariant, i.e., $d_a(x, y) = d_a(x + z, y + z)$ for all $x, y, z \in I^\omega$ with $x + z, y + z \in I^\omega$.

We denote by the symbol π the standard product probability measure on I^ω.

Lemma 5.8. *The Borel measure μ of Theorem 5.6 is unique and $\mu = \pi$.*

Proof. Since d_a is translation-invariant, so is μ over all open subsets of I^ω. We now consider a covering of I^n with m^n isometric n-cubes with non-overlapping interiors. We denote by $\mathcal{C}_{mn}$ the collection of m^n open cylinders in I^ω over the interiors of those n-cubes.

We will prove that

$$\mu(K) = \frac{1}{m^n} = \pi(K) \tag{5.2}$$

for all cylinders $K \in \mathcal{C}_{mn}$. Since μ is translation-invariant over all open subsets of I^ω, we have $\mu(K_1) = \mu(K_2)$ for all cylinders $K_1, K_2 \in \mathcal{C}_{mn}$. On the other hand, it follows from Proposition 5.7 that the boundary $\partial(\bigcup \mathcal{C}_{mn})$ of $\bigcup \mathcal{C}_{mn}$ is a finite union of thin subsets of I^ω. Thus, by Theorem 5.6 (ii), we conclude that $\mu(\partial(\bigcup \mathcal{C}_{mn})) = 0$.

Since the collection of all open cylinders of $\bigcup_{m,n=1}^{\infty} \mathcal{C}_{mn}$ generates the Borel σ-algebra over I^ω, by well-known facts, (5.2) implies $\mu = \pi$. □

J. Mycielski proved the conjecture of Ulam affirmatively under the additional assumption that the sets are open.

Theorem 5.9. *Let U_1 and U_2 be open subsets of the Hilbert cube I^ω. If U_1 is d_a-isometric to U_2, then $\pi(U_1) = \pi(U_2)$.*

Proof. The assertion of this theorem can easily be proved by using Theorem 5.6 and Lemma 5.8. □

As we have seen in this section, J. Mycielski has positively proved Ulam's conjecture under the additional assumption that the related sets are open. And he asked whether one could prove Ulam's conjecture by assuming that the related sets are closed instead of open.

5.4 Fickett's Partial Solution

In 1982, J. W. Fickett succeeded in partially proving Ulam's conjecture by proving the following in a different way from J. Mycielski: if the a_i's tend to 0 fast enough so that

$$\frac{a_{i+1}^{1/2^{i+1}}}{a_i} \to 0 \quad \text{as} \quad i \to \infty, \tag{5.3}$$

then any two Borel subsets of the Hilbert cube I^ω which are d_a-isometric have the same standard product probability measure.

For any subset E of $\mathbb{R}^n$, we denote by $H(E)$ the smallest *flat* containing E, i.e.,

$$\begin{aligned} H(E) = \big\{ &\lambda_0 p_0 + \lambda_1 p_1 + \cdots + \lambda_n p_n : p_0, p_1, \ldots, p_n \in E; \\ &\lambda_0, \lambda_1, \ldots, \lambda_n \in \mathbb{R} \text{ and } \lambda_0 + \lambda_1 + \cdots + \lambda_n = 1 \big\}. \end{aligned}$$

We will write $H(p_0, p_1, \ldots, p_k)$ instead of $H(\{p_0, p_1, \ldots, p_k\})$.

The points $p_0, p_1, \ldots, p_k$ of $\mathbb{R}^n$ are said to be *independent* if $H(p_0, p_1, \ldots, p_k)$ is k-dimensional. For example, three points are independent if and only if they are not collinear.

We use $d_e(\cdot, \cdot)$ to denote the usual Euclidean distance for $\mathbb{R}^n$. It is easy to check that if $p_0, p_1, \ldots, p_n$ are independent in $\mathbb{R}^n$ and $d_0, d_1, \ldots, d_n \geq 0$, then the equations $d_e(x, p_i) = d_i$, for all $i \in \{0, 1, \ldots, n\}$, have at most one solution $x \in \mathbb{R}^n$. The following is also easy to prove.

Proposition 5.10. *If $p_0, p_1, \ldots, p_{n-2}$ are independent in $\mathbb{R}^n$ and $d_0, d_1, \ldots, d_{n-2} \geq 0$, then the subset S of $\mathbb{R}^n$ defined by*

$$S = \big\{ x \in \mathbb{R}^n : d_e(x, p_i) = d_i \text{ for } i \in \{0, 1, \ldots, n-2\} \big\}$$

is a (possibly degenerate) circle with center in $H(p_0, p_1, \ldots, p_{n-2})$. If S is nondegenerate, then $H(S)$ is perpendicular to $H(p_0, p_1, \ldots, p_{n-2})$.

The following lemma is a special case of Lemma 5.12. The latter is the main tool for the proof of Theorem 5.13.

Lemma 5.11. *Let S and c be a circle and a point in $\mathbb{R}^3$, respectively. Assume that the perpendicular projection of c on the plane of S lies outside of S or on S. Let d be given such that $0 < d \leq 1$ and the ball with center c and radius d intersects S. For any $0 < \varepsilon \leq 1$ and any point p on S with $|d - d_e(p, c)| \leq \varepsilon$, there exists a point q on S with $d_e(c, q) = d$ and $d_e(p, q) \leq 2\sqrt{\varepsilon}$.*

Proof. Consider the ball with center c and radius d and also the spheres with center c and radii $d - \varepsilon$ and $d + \varepsilon$, and intersect them with the plane of S. Then the ball and spheres become a circle S' and an annulus about S', respectively. We note that S' has its center c' outside of S or on S. Let q be one of the two points of $S \cap S'$ closer to p. The worst case is when p lies on the outer boundary of the annulus and the line segment $\overline{qp}$ is tangent to S'. In this case, we get the final result by applying the Pythagorean theorem to the triangle Δcpq. □

Lemma 5.12. *Assume that $p_0, p_1, \ldots, p_{n-1}$ are independent points of $\mathbb{R}^n$. Define $H = H(p_0, p_1, \ldots, p_{n-2})$ and let p_n be any point of $\mathbb{R}^n$ that is no further away from H as p_{n-1}. Define $d_i = d_e(p_n, p_i)$ for all $i \in \{0, 1, \ldots, n-2\}$. Assume that $d_i \leq 1$ for $i \in \{0, 1, \ldots, n-2\}$ and that d_{n-1} and $0 < \varepsilon \leq 1$ are given with $|d_{n-1} - d_e(p_n, p_{n-1})| \leq \varepsilon$. If there exists a point q with $d_e(q, p_i) = d_i$ for all $i \in \{0, 1, \ldots, n-1\}$, then there exists such a q with $d_e(p_n, q) \leq 2\sqrt{\varepsilon}$.*

Proof. Due to Proposition 5.10, the set

$$S = \big\{x \in \mathbb{R}^n : d_e(x, p_i) = d_i \text{ for each } i \in \{0, 1, \ldots, n-2\}\big\}$$

is a circle with radius r and center in H.

If $r = 0$, then $p_n = q$, so we are done. So let us assume that $r > 0$. Then, by Proposition 5.10, H is perpendicular to S and passes through the center of circle S. Let K be a three-dimensional flat containing S and p_{n-1}. Then $H \cap K$ is a line through the center of S, perpendicular to the plane of S. Moreover, p_{n-1} is at least r from this line, and p_n is on S. The ball of radius d_{n-1} about p_{n-1} intersects S.

So we may apply Lemma 5.11 with S, p, c, d, and q there equal to S, p_n, p_{n-1}, d_{n-1}, and q here, respectively, to get the desired result. □

Let (X, d_1) and (Y, d_2) be metric spaces, and let $\delta > 0$. We remember that a function $g : X \to Y$ is called a δ-isometry if

$$\big|d_2\big(g(x), g(y)\big) - d_1(x, y)\big| \leq \delta$$

for all $x, y \in X$. That is, a δ-isometry is a function that preserves distances within δ.

First we describe a method for converting a δ-isometry, which maps a bounded subset of $\mathbb{R}^n$ into $\mathbb{R}^n$, into an isometry: let S be a bounded subset of $\mathbb{R}^n$ and let $g : S \to \mathbb{R}^n$ be a δ-isometry for some $\delta > 0$.

Step 1. We extend the δ-isometry g to $\overline{S}$, where $\overline{S}$ denotes the closure of S. We define a function $g_1 : \overline{S} \to \mathbb{R}^n$ by

$$g_1(x) \in \begin{cases} \bigcap\limits_{i=1}^{\infty} \overline{g(\{y \in S : d_e(x,y) < \frac{1}{i}\})} & (\text{for } x \in \overline{S} \setminus S), \\ \{g(x)\} & (\text{for } x \in S). \end{cases}$$

Then we can easily show that g_1 is also a δ-isometry.

Step 2. Assume that $H(S)$ is k-dimensional. We select $s_0, s_1, \ldots, s_k \in \overline{S}$ such that

$$\begin{aligned} &d_e(s_0, s_1) = d(\overline{S}), \\ &d_e\big(s_i, H(s_0, s_1, \ldots, s_{i-1})\big) = \sup\big\{d_e(s, H(s_0, s_1, \ldots, s_{i-1})) : s \in S\big\} \end{aligned} \tag{5.4}$$

for all $i \in \{2, 3, \ldots, k\}$. We note that the conditions in (5.4) force $s_0, s_1, \ldots, s_k$ to be independent.

Step 3. We define inductively

$$\begin{aligned} f(s_0) &= g_1(s_0), \\ f(s_i) &= \text{a point as close as possible to } g_1(s_i) \text{ satisfying} \\ &\qquad d_e(f(s_i), f(s_j)) = d_e(s_i, s_j) \text{ for } 0 \leq j \leq i \leq k. \end{aligned} \tag{5.5}$$

And finally, for any $s \in \overline{S}$, we define

$$\begin{aligned} f(s) &= \text{the unique point in } H(f(s_0), f(s_1), \ldots, f(s_k)) \text{ satisfying} \\ &\qquad d_e(f(s), f(s_i)) = d_e(s, s_i) \text{ for } 0 \leq i \leq k. \end{aligned} \tag{5.6}$$

We remember that we use the symbol $d(S)$ to denote the diameter of any subset S of $\mathbb{R}^n$, i.e., $d(S) = \sup\{d_e(x,y) : x, y \in S\}$.

For any $\delta \geq 0$, we define

$$K_0(\delta) = K_1(\delta) = \delta, \quad K_2(\delta) = 3(3\delta)^{1/2}, \quad \text{and} \quad K_i(\delta) = 27\delta^{2^{1-i}} \tag{5.7}$$

for each integer $i \geq 3$.

Theorem 5.13. *Let S be a bounded subset of $\mathbb{R}^n$ and $g : S \to \mathbb{R}^n$ a δ-isometry, where $0 \leq \delta \leq 1$ and $0 \leq 3K_n(\frac{\delta}{d(S)}) \leq 1$. Then the isometry $f : S \to \mathbb{R}^n$ gotten by applying the above construction to g satisfies*

$$\sup\big\{d_e(f(s), g(s)) : s \in S\big\} \leq K_{n+1}\bigg(\frac{\delta}{d(S)}\bigg) d(S).$$

Proof. The general case can be reduced to the case $d(S) = 1$ by a homothety argument. Apply the above construction to g, and let $g_1, s_0, s_1, \ldots, s_k$ and f be as given there.

We apply induction on m to prove that if $t_0, t_1, \ldots, t_m \in \overline{S}$, with $t_0, t_1, \ldots, t_{m-1}$ independent, satisfy

$$d_e\big(t_i, H(t_0, t_1, \ldots, t_{i-1})\big) \geq d_e\big(t_j, H(t_0, t_1, \ldots, t_{i-1})\big), \tag{5.8}$$

for all integers $1 \leq i \leq j \leq m$, and if $h : \{t_0, t_1, \ldots, t_m\} \to \mathbb{R}^n$ is defined inductively by

$h(t_i)$ is a point as close as possible to $g_1(t_i)$ satisfying
$d_e(h(t_i), h(t_j)) = d_e(t_i, t_j)$ for all integers $0 \leq j \leq i \leq m$,

then

$$d_e\big(h(t_i), g_1(t_i)\big) \leq K_m(\delta)$$

for all $i \in \{0, 1, \ldots, m\}$.

This assertion is obviously true for $m \in \{0, 1\}$. For some integer $m \geq 2$, we assume the truth of the assertion for m-point sets, and let $t_0, t_1, \ldots, t_m$ and h be as described. Then each of $\{t_0, \ldots, t_{m-2}, t_{m-1}\}$ and $\{t_0, t_1, \ldots, t_{m-2}, t_m\}$ satisfies an m-point version of (5.8).

Let us define

$$h_1(t_i) = \begin{cases} h(t_i) & (\text{for } 0 \leq i \leq m-1), \\ \text{a point as close as possible to } g_1(t_m) & \\ \text{satisfying } d_e(h_1(t_m), h_1(t_i)) = d_e(t_m, t_i) & (\text{for } i = m). \\ \text{for } 0 \leq i \leq m-2 & \end{cases}$$

It can easily be proved by induction that $h_1(t_m) \in H(h(t_0), h(t_1), \ldots, h(t_m))$.

By the induction hypothesis, we have

$$d_e\big(h_1(t_i), g_1(t_i)\big) \leq K_{m-1}(\delta)$$

for any $i \in \{0, 1, \ldots, m\}$. Hence, we get

$$\big|d_e\big(h_1(t_{m-1}), h_1(t_m)\big) - d_e(t_{m-1}, t_m)\big| \leq \delta + 2K_{m-1}(\delta) \leq 3K_{m-1}(\delta).$$

Thus, we can apply Lemma 5.12 with $p_0, p_1, \ldots, p_m, d_{m-1}, \varepsilon$, and $\mathbb{R}^m$ there equal to $h_1(t_0), h_1(t_1), \ldots, h_1(t_m), d_e(t_m, t_{m-1}), 3K_{m-1}(\delta)$, and $H(h(t_0), \ldots, h(t_m))$

to get a $q \in H(h(t_0), h(t_1), \dots, h(t_m))$ with $d_e(q, h(t_i)) = d_e(t_m, t_i)$ for all $i \in \{0, 1, \dots, m-2\}$ and $d_e(q, h(t_m)) \leq 2(3K_{m-1}(\delta))^{1/2}$. Thus, we have

$$\begin{aligned} d_e\big(g_1(t_m), h(t_m)\big) &\leq d_e\big(g_1(t_m), q\big) \\ &\leq d_e\big(g_1(t_m), h_1(t_m)\big) + d_e\big(h_1(t_m), q\big) \\ &\leq K_{m-1}(\delta) + 2\big(3K_{m-1}(\delta)\big)^{1/2} \\ &\leq K_m(\delta), \end{aligned}$$

which completes the proof of our first assertion.

Finally, let $s \in S$ be arbitrary. By (5.4), (5.5), and (5.6), we can apply the proceeding with $m = k+1$, $t_0 = s_0, t_1 = s_1, \dots, t_k = s_k$, $t_m = s$, and $h = f$ to conclude that

$$d_e\big(f(s), g_1(s)\big) \leq K_{k+1}(\delta) \leq K_{n+1}(\delta)$$

for all $s \in S$. □

In the following theorem, a measure μ defined on the Borel σ-algebra of subsets of the Hilbert cube I^ω is said to be d_a-invariant if d_a-isometric Borel sets have the same μ-measure.

Theorem 5.14. *Let $\{a_i\}_{i \in \mathbb{N}_0}$ be a sequence of positive real numbers that satisfies the condition* (5.3). *Then the standard product probability measure π on I^ω is d_a-invariant.*

Proof. The proof uses Theorem 5.13 and a modification of the reduction of Ulam's conjecture proposed by J. Mycielski in [17, Theorem 5].

Assume that $a = \{a_i\}_{i \in \mathbb{N}_0}$ is a sequence of positive real numbers that satisfies the condition (5.3). We set

$$r_n = \left(\sum_{i=n}^{\infty} a_i^2 \right)^{1/2} \quad \text{and} \quad \varepsilon_n = 27 r_n^{1/2^n} r_0^{1-1/2^n}$$

for all $n \in \mathbb{N}_0$. We note that $\varepsilon_n = r_0 K_{n+1}(\frac{r_n}{r_0})$, where K_{n+1} is defined in (5.7).

Let us define the parallelepiped P_n in $\mathbb{R}^n$ by

$$P_n = [0, a_0] \times [0, a_1] \times \cdots \times [0, a_{n-1}].$$

Let

$$P_n' = \big\{ x \in \mathbb{R}^n : d_e(x, P_n) \leq \varepsilon_n \big\},$$

where d_e is the usual Euclidean distance for $\mathbb{R}^n$, and

$$\delta_n = \frac{\mu_n(P_n')}{\mu_n(P_n)} - 1,$$

where μ_n denotes the n-dimensional Lebesgue measure.

We assert that

$$\delta_n \to 0 \quad \text{as} \quad n \to \infty. \tag{5.9}$$

Indeed, it easily follows from (5.3) that

$$a_n \leq \frac{1}{2^n} a_{n-1} \tag{5.10}$$

and hence

$$\begin{aligned} r_n^2 &= \sum_{i=n}^{\infty} a_i^2 \\ &= a_n^2 + a_{n+1}^2 + a_{n+2}^2 + \cdots \\ &\leq a_n^2 + \frac{1}{4^{n+1}} a_n^2 + \frac{1}{4^{n+2}} a_{n+1}^2 + \cdots \\ &\leq a_n^2 + \frac{1}{4^{n+1}} a_n^2 + \frac{1}{4^{n+2}} \frac{1}{4^{n+1}} a_n^2 + \cdots \\ &\leq 4a_n^2 \end{aligned}$$

for all sufficiently large integers n. Hence, it follows from (5.3) that

$$\begin{aligned} \left(\frac{\varepsilon_n}{a_{n-1}}\right)^2 &= \frac{27^2 r_n^{2/2^n} r_0^{2-2/2^n}}{a_{n-1}^2} \\ &\leq \frac{27^2 (4a_n^2)^{1/2^n} r_0^{2-2/2^n}}{a_{n-1}^2} \\ &= 27^2 4^{1/2^n} r_0^{2-2/2^n} \frac{a_n^{1/2^{n-1}}}{a_{n-1}^2} \\ &\to 0 \end{aligned} \tag{5.11}$$

as $n \to \infty$. Thus, there exists an $n_0 \in \mathbb{N}$ such that

$$\frac{\varepsilon_n}{a_{n-1}} \leq \frac{1}{2} \quad \text{or} \quad 2\frac{\varepsilon_n}{a_i} \leq \frac{a_{n-1}}{a_i}$$

for all integers $n > n_0$.

Therefore, we have

$$\begin{aligned}
\frac{\mu_n(P'_n)}{\mu_n(P_n)} &\leq \frac{(a_0+2\varepsilon_n)(a_1+2\varepsilon_n)\cdots(a_{n-1}+2\varepsilon_n)}{a_0a_1\cdots a_{n-1}} \\
&= \prod_{i=0}^{n-1}\left(1+2\frac{\varepsilon_n}{a_i}\right) \\
&\leq \left(1+2\frac{\varepsilon_n}{a_{n-1}}\right)\prod_{i=0}^{n_0}\left(1+2\frac{\varepsilon_n}{a_i}\right)\prod_{i=n_0+1}^{n-2}\left(1+\frac{a_{n-1}}{a_i}\right)
\end{aligned}$$

for all integers $n > n_0$. It follows from (5.10) that

$$\begin{aligned}
\frac{a_{n-1}}{a_i} &= \frac{a_{n-1}}{a_{n-2}}\cdot\frac{a_{n-2}}{a_{n-3}}\cdots\frac{a_{i+1}}{a_i} \\
&\leq \frac{1}{2^{n-1}}\cdot\frac{1}{2^{n-2}}\cdots\frac{1}{2^{i+1}} \\
&= \frac{1}{2^{((n-1)n-i(i+1))/2}} \\
&\leq \frac{1}{2^{n-1}}
\end{aligned}$$

for all $i \in \{n_0+1, n_0+2, \ldots, n-2\}$. Hence, we have

$$1+\frac{a_{n-1}}{a_i} \leq 1+2^{1-n},$$

for all $i \in \{n_0+1, n_0+2, \ldots, n-2\}$, and

$$\frac{\mu_n(P'_n)}{\mu_n(P_n)} \leq \left(1+2\frac{\varepsilon_n}{a_{n-1}}\right)\left(\prod_{i=0}^{n_0}\left(1+2\frac{\varepsilon_n}{a_i}\right)\right)(1+2^{1-n})^{n-n_0-2}$$

for all integers $n > n_0$, which tends to 1 as n tends to ∞, and thus, (5.9) holds.

We define a function $\omega_n : I^\omega \to P_n$ by

$$\omega_n(x) = (a_0x_0, a_1x_1, \ldots, a_{n-1}x_{n-1})$$

and

$$\nu_n(E) = \frac{1}{a_0a_1\cdots a_{n-1}}\mu_n(E)$$

for any Borel set $E \subset P_n$.

Furthermore, we define

$$E^{(t)} = \begin{cases} \{x \in I^{\omega} : d_a(x, E) \le t\} & (\text{for } E \subset I^{\omega}), \\ \{x \in P_n : d_e(x, E) \le t\} & (\text{for } E \subset P_n) \end{cases}$$

for any $t \ge 0$. Thus it holds that

$$\big(\omega_n(E)\big)^{(t)} = \omega_n\big(E^{(t)}\big) \tag{5.12}$$

for $E \subset I^{\omega}$.

We assert that

$$\nu_n\Big(\omega_n\big(S^{(\varepsilon_n)}\big)\Big) \to \pi(S) \quad \text{as} \quad n \to \infty \tag{5.13}$$

for all compact subsets S of I^{ω}. Let $\eta > 0$ be given. Since π is regular, we can select $t > 0$ such that

$$\pi\big(S^{(t)}\big) < \pi(S) + \frac{1}{2}\eta.$$

Since π is a product measure, there exists an $N \in \mathbb{N}$ such that

$$\nu_n\Big(\omega_n\big(S^{(t)}\big)\Big) < \pi\big(S^{(t)}\big) + \frac{1}{2}\eta < \pi(S) + \eta$$

for all $n > N$. The condition (5.3) or (5.11) implies that $\varepsilon_n \to 0$ as $n \to \infty$. Thus, we choose an $M > N$ such that $\varepsilon_n < t$ for $n > M$. Then we have

$$\pi(S) < \nu_n\Big(\omega_n\big(S^{(\varepsilon_n)}\big)\Big) < \nu_n\Big(\omega_n\big(S^{(t)}\big)\Big) < \pi(S) + \eta$$

for all $n > M$, which proves (5.13).

Now we fix a compact subset S of I^{ω} and a d_a-isometry $f : S \to I^{\omega}$. Let $q_n : \omega_n(S) \to S$ be any function satisfying $q_n(x) \in \omega_n^{-1}(x)$ for all $x \in \omega_n(S)$. Moreover, we define a function $F_n : \omega_n(S) \to P_n$ by

$$F_n = \omega_n \circ f \circ q_n.$$

Then

$$F_n\big(\omega_n(S)\big) \subset \omega_n\big(f(S)\big). \tag{5.14}$$

Since

$$d_e(x, y) \le d_a\big(q_n(x), q_n(y)\big) \le d_e(x, y) + r_n$$

and

$$\begin{aligned} d_a\big(q_n(x), q_n(y)\big) - r_n &= d_a\big(f(q_n(x)), f(q_n(y))\big) - r_n \\ &\le d_e\big(F_n(x), F_n(y)\big) \\ &\le d_a\big(f(q_n(x)), f(q_n(y))\big) \\ &= d_a\big(q_n(x), q_n(y)\big) \end{aligned}$$

for all $x, y \in \omega_n(S)$, it holds that F_n is an r_n-isometry. Hence, due to Theorem 5.13, there exists an isometry $f_n : \omega_n(S) \to \mathbb{R}^n$ within ε_n of F_n. Therefore, it holds that $f_n(\omega_n(S)) \subset P_n'$.

Since

$$\begin{aligned} \mu_n\big(f_n(\omega_n(S)) \cap P_n\big) &\ge \mu_n\big(f_n(\omega_n(S))\big) - \mu_n\big(P_n' \setminus P_n\big) \\ &\ge \mu_n\big(\omega_n(S)\big) - \delta_n \mu_n(P_n), \end{aligned}$$

we have

$$\mu_n\Big(\big(F_n(\omega_n(S))\big)^{(\varepsilon_n)}\Big) \ge \mu_n\big(\omega_n(S)\big) - \delta_n \mu_n(P_n)$$

or

$$\nu_n\Big(\big(F_n(\omega_n(S))\big)^{(\varepsilon_n)}\Big) \ge \nu_n\big(\omega_n(S)\big) - \delta_n. \tag{5.15}$$

Thus, given $\varepsilon > 0$, we select an $N \in \mathbb{N}$ such that

$$\begin{aligned} \pi\big(f(S)\big) + \varepsilon &\ge \nu_n\Big(\omega_n\big(f(S)^{(\varepsilon_n)}\big)\Big) && \text{by (5.13)} \\ &= \nu_n\Big(\big(\omega_n(f(S))\big)^{(\varepsilon_n)}\Big) && \text{by (5.12)} \\ &\ge \nu_n\Big(\big(F_n(\omega_n(S))\big)^{(\varepsilon_n)}\Big) && \text{by (5.14)} \\ &\ge \nu_n\big(\omega_n(S)\big) - \delta_n && \text{by (5.15)} \\ &\ge \pi(S) - \delta_n \end{aligned}$$

for all $n \ge N$. It follows from (5.9) that $\delta_n \to 0$ as $n \to \infty$. Hence, we have $\pi(f(S)) \ge \pi(S)$. By its symmetric property, it holds that $\pi(f(S)) = \pi(S)$. Thus, d_a-isometric compact subsets of I^ω have the same π-measure, and so, by regularity of π, π is d_a-invariant. □

5.5 Jung and Kim's Partial Solution

S.-M. Jung and E. Kim proved that Ulam's conjecture is true if the sequence $a = \{a_i\}_{i \in \mathbb{N}}$ of positive real numbers is monotone decreasing and satisfies the condition

$$a_{i+1} = o\bigg(\frac{a_i}{\sqrt{i}}\bigg) \quad \text{as} \quad i \to \infty.$$

In other words, they proved that Ulam's conjecture is true if the sequence $a = \{a_i\}_{i\in\mathbb{N}}$ satisfies the equivalent condition

$$\frac{\sqrt{i}\,a_{i+1}}{a_i} \to 0 \quad \text{as} \quad i \to \infty. \tag{5.16}$$

Obviously the last condition is much weaker than the condition (5.3) of Fickett.

In this section, we assume that $a = \{a_i\}_{i\in\mathbb{N}}$ is a monotone decreasing sequence of positive real numbers satisfying the condition (5.16) and we define a sequence $\{\delta_i\}_{i\in\mathbb{N}}$ by

$$\delta_i^2 = \sum_{j=i+1}^{\infty} a_j^2 + a_{i+1}^2 \tag{5.17}$$

for any $i \in \mathbb{N}$. We set $I = [0, 1]$ and

$$\mathcal{G} = \big\{G \subset M_a : \text{there exist an } F \in \mathcal{F} \text{ and a } d_a\text{-isometry of } F \text{ onto } G\big\}, \tag{5.18}$$

where $\mathcal{F}$ is the set of all non-degenerate basic cylinders in I^ω, i.e.,

$$\mathcal{F} = \left\{\prod_{i=1}^{\infty} I_i : I_i \text{ is a non-degenerate closed interval in } I;\right.$$
$$\left. I_i \neq I \text{ for at most finitely many } i\right\}.$$

We recall that we use the symbol $d_a(E)$ to denote the diameter of any subset E of M_a, i.e., $d(E) = \sup\{d_a(x, y) : x, y \in E\}$. Moreover, we set

$$\mathcal{U} = \bigcup_{i=1}^{\infty} \{G \in \mathcal{G} : d_a(G) = \delta_i\} \cup \{\emptyset\} \tag{5.19}$$

and define

$$\mathcal{F}_\delta = \big\{F \in \mathcal{F} : d_a(F) < \delta\big\},$$
$$\mathcal{G}_\delta = \big\{G \in \mathcal{G} : d_a(G) < \delta\big\},$$
$$\mathcal{U}_\delta = \big\{U \in \mathcal{U} : d_a(U) < \delta\big\}$$

for all $\delta > 0$.

For a fixed compact subset K of a metric space (X, d), let $E(K, \delta)$ be the entropy that denotes the least number of sets of diameter $< \delta$ necessary to cover K. With this notation, J. Mycielski has introduced the following pre-measure:

$$h(G) = \frac{1}{E(K, d(G))}$$

to construct some special Hausdorff measures (see Sect. 5.3).

We apply this definition in the modified form

$$\tau(G) = \frac{1}{E'(I^\omega, d_a(G))} \quad \text{and} \quad \tau(\emptyset) = 0, \tag{5.20}$$

where $E'(I^\omega, \delta)$ denotes the least number of sets in $\mathcal{U}_\delta$ necessary to cover I^ω and where τ is defined on the collection of all cylinders in $\bigcup_{k=1}^{\infty} \mathcal{U}_{\delta_k}$.

Let J be a (closed) basic cylinder from $\mathcal{F}$ and let K be an arbitrary cylinder in $\mathcal{G}$ such that there exists a d_a-isometry f of J onto K. Since J is a compact subset of M_a as a closed subset of a compact set I^ω, Theorem 1.23 implies that K is also a compact subset of M_a as the continuous image of a compact set J. Moreover, by Theorem 1.43, we conclude that K is a closed subset of M_a as a compact subset of the Hausdorff space M_a. Hence, we come to an important consequence.

Remark 5.15.

(i) Every cylinder in $\mathcal{G}$ is closed in I^ω.

(ii) $\mathcal{G}$ and $\mathcal{U}$ are closed under the actions of d_a-isometries.

It is worth noting that τ does not depend on the shape of sets, but it depends on the diameter of the sets only.

According to Theorem 3.24, an outer measure μ^* may be constructed by applying the "Method II" from the pre-measure τ as we see in the following:

$$\mu^*(C) = \sup\left\{\mu_k^*(C) : k \in \mathbb{N}\right\}$$

for all subsets C of M_a, where

$$\mu_k^*(C) = \inf\left\{\sum_{i=1}^{\infty} \tau(U_i) : \{U_i\}_{i\in\mathbb{N}} \subset \mathcal{U}_{\delta_k} \text{ is a covering of } C\right\}.$$

Remark 5.16. According to Theorems 3.26 and 3.30, we remark that

(i) The outer measure μ^* is metric.

(ii) Each Borel set in I^ω is μ^*-measurable.

(iii) The translation invariance of d_a implies that μ^* on I^ω is also invariant under translation.

For any cylinder $G \in \mathcal{G}$, we denote by $|G|_i$, $i \in \mathbb{N}$, the (Euclidean) length of the ith edge of G and by $\mathrm{vol}(G) = \prod_{i=1}^{\infty} |G|_i$ the *volume* of G.

Which of the cylinders $G \in \mathcal{G}$ of a given diameter D has the largest volume?

The following two lemmas provide a partial answer to the above question.

Lemma 5.17. *Let D be a sufficiently small positive number, $m \in \mathbb{N}$, and $G \in \mathcal{G}$ any cylinder. Then*

$$|G|_i = \begin{cases} \dfrac{D}{a_i\sqrt{m}} & (\textit{for } i \in \{1, 2, \ldots, m\}), \\ 1 & (\textit{for } i > m) \end{cases} \tag{5.21}$$

if and only if

$$\begin{aligned} \mathrm{vol}(G) = \max \Bigg\{ & \mathrm{vol}(G') : G' \in \mathcal{G} \textit{ is a cylinder satisfying} \\ & \sum_{i=1}^{m} a_i^2 |G'|_i^2 = D^2 \textit{ and } |G'|_i = 1 \textit{ for } i > m \Bigg\}. \end{aligned} \tag{5.22}$$

Proof. There exist an $F \in \mathcal{F}$ and a d_a-isometry of F onto G and, according to Remark 6.5, it holds that $\mathrm{vol}(F) = \mathrm{vol}(G)$. Hence, it suffices to prove our lemma for the basic cylinders in $\mathcal{F}$.

First, under the assumption (5.21), we prove that if $G' \in \mathcal{F}$ and $\sum_{i=1}^{m} a_i^2 |G'|_i^2 = D^2$ (we temporarily neglect the condition, $|G'|_i = 1$ for $i > m$), then

$$\prod_{i=1}^{m} |G'|_i \leq \prod_{i=1}^{m} |G|_i \tag{5.23}$$

for all $m \in \mathbb{N}$ and $D > 0$. This assertion is true for $m = 1$ and for any sufficiently small $D > 0$. Now, assume that our assertion is valid for $m = q - 1$ ($q \geq 2$), i.e., we assume that if $D > 0$, $G' \in \mathcal{F}$, and $\sum_{i=1}^{q-1} a_i^2 |G'|_i^2 = D^2$, then

$$\prod_{i=1}^{q-1} |G'|_i \leq \prod_{i=1}^{q-1} |G|_i = \frac{1}{a_1 a_2 \cdots a_{q-1}} \left(\frac{D}{\sqrt{q-1}} \right)^{q-1}. \tag{5.24}$$

Let x be a sufficiently small positive number and let $G' \in \mathcal{F}$ be a basic cylinder satisfying $\sum_{i=1}^{q} a_i^2 |G'|_i^2 = D^2$ and $|G'|_q = x$. Then, $\sum_{i=1}^{q-1} a_i^2 |G'|_i^2 = D^2 - a_q^2 x^2 > 0$ (we except the degenerate case, $|G'|_i = 0$ for some i, from our consideration).

Since we assumed that inequality (5.24) holds for all sufficiently small $D > 0$, if we replace D with $\sqrt{D^2 - a_q^2 x^2}$ in (5.24), then

$$\prod_{i=1}^{q-1} |G'|_i \leq \frac{1}{a_1 a_2 \cdots a_{q-1}} \left(\frac{D^2 - a_q^2 x^2}{q-1} \right)^{(q-1)/2}$$

and

$$\prod_{i=1}^{q} |G'|_i = x \prod_{i=1}^{q-1} |G'|_i \leq \frac{x}{a_1 a_2 \cdots a_{q-1}} \left(\frac{D^2 - a_q^2 x^2}{q-1} \right)^{(q-1)/2} =: v(x). \quad (5.25)$$

If we differentiate $v(x)$, then

$$v'(x) = \frac{1}{a_1 a_2 \cdots a_{q-1}} \left(\frac{D^2 - a_q^2 x^2}{q-1} \right)^{(q-3)/2} \cdot \frac{D^2 - q a_q^2 x^2}{q-1}.$$

Since $D^2 - a_q^2 x^2 > 0$, $v(x)$ has at $x = x_0 := \frac{D}{a_q \sqrt{q}}$ its maximum

$$v(x_0) = \frac{1}{a_1 a_2 \cdots a_q} \left(\frac{D}{\sqrt{q}} \right)^q = \prod_{i=1}^{q} \frac{D}{a_i \sqrt{q}} = \prod_{i=1}^{q} |G|_i. \quad (5.26)$$

In view of (5.25) and (5.26), we proved that

$$\prod_{i=1}^{q} |G'|_i \leq v(x) \leq v(x_0) = \prod_{i=1}^{q} |G|_i$$

for all basic cylinders $G' \in \mathcal{F}$ with $\sum_{i=1}^{q} a_q^2 |G'|_i^2 = D^2$ and for $m = q$. Indeed, we proved the validity of inequality (5.23) for all $m \in \mathbb{N}$. Considering

$$\mathrm{vol}(G') = \prod_{i=1}^{\infty} |G'|_i \leq \prod_{i=1}^{m} |G'|_i \leq \prod_{i=1}^{m} |G|_i = \prod_{i=1}^{\infty} |G|_i = \mathrm{vol}(G)$$

and

$$\sum_{i=1}^{m} a_i^2 |G|_i^2 = D^2,$$

we conclude that (5.21) implies (5.22).

Finally, we will prove that (5.22) implies (5.21). Obviously, (5.22) implies (5.21) for $m = 1$. We now assume that $m > 1$ and $G \in \mathcal{G}$ is a cylinder with $|G|_i = 1$ for all $i > m$. Then we have

$$\mathrm{vol}(G) = \prod_{i=1}^{\infty} |G|_i = \prod_{i=1}^{m} |G|_i = \prod_{i=1}^{m} x_i =: f(x_1, x_2, \ldots, x_m),$$

where we temporarily set $x_i := |G|_i > 0$ for every $i \in \{1, 2, \ldots, m\}$.

It is to prove that $x_i = \frac{D}{a_i \sqrt{m}}$ holds for each $i \in \{1, 2, \ldots, m\}$ under the condition

$$g(x_1, x_2, \ldots, x_m) := \sum_{i=1}^{m} a_i^2 x_i^2 = D^2, \tag{5.27}$$

which is given in (5.22). In other words, we will prove that if G is a cylinder satisfying the condition (5.22), then G has the form of (5.21). We apply the method of Lagrange multipliers to maximize the value of $f(x_1, x_2, \ldots, x_m)$ subject to the condition (5.27) (ref. [22]). We introduce a new variable λ and define the Lagrange function by

$$\mathcal{L}(x_1, x_2, \ldots, x_m, \lambda) := f(x_1, x_2, \ldots, x_m) - \lambda\big(g(x_1, x_2, \ldots, x_m) - D^2\big)$$

and solve

$$\frac{\partial}{\partial x_i}\mathcal{L}(x_1, x_2, \ldots, x_m, \lambda) = 0 \quad \text{and} \quad \frac{\partial}{\partial \lambda}\mathcal{L}(x_1, x_2, \ldots, x_m, \lambda) = 0$$

for all $i \in \{1, 2, \ldots, m\}$.

Thus, we have

$$\frac{x_1 x_2 \cdots x_m}{x_i} - 2\lambda a_i^2 x_i = 0 \quad \text{and} \quad \sum_{j=1}^{m} a_j^2 x_j^2 = D^2 \tag{5.28}$$

for any $i \in \{1, 2, \ldots, m\}$. From the first equalities of (5.28), we get

$$\lambda = \frac{x_1 x_2 \cdots x_m}{2 a_i^2 x_i^2} \quad \text{or} \quad a_1^2 x_1^2 = a_2^2 x_2^2 = \cdots = a_m^2 x_m^2$$

for any $i \in \{1, 2, \ldots, m\}$. By the second equality of (5.28), we obtain

$$|G|_i = x_i = \frac{D}{a_i\sqrt{m}}$$

for all $i \in \{1, 2, \ldots, m\}$, which implies that G has the form of (5.21). □

Lemma 5.18. *Let $a = \{a_i\}_{i\in\mathbb{N}}$ be a monotone decreasing sequence of positive real numbers satisfying the condition* (5.16). *Let $j \in \mathbb{N}$ be sufficiently large. Assume that U is a cylinder in $\mathcal{U}$ with $d_a(U)^2 = \sum_{i=j+1}^{\infty} a_i^2 + a_{j+1}^2$ and*

$$|U|_i = \begin{cases} \dfrac{\sqrt{2}\,a_{j+1}}{a_i\sqrt{j+1}} & (\text{for } i \in \{1, 2, \ldots, j+1\}), \\ 1 & (\text{for } i > j+1). \end{cases}$$

Then

$$\mathrm{vol}(U) \geq \frac{1}{2} \sup\left\{\mathrm{vol}(U') : U' \in \mathcal{G}_{d_a(U)}\right\}.$$

Proof. In view of the definition of $\mathcal{U}$ and Remark 6.5, it suffices to prove our lemma for the intervals in $\mathcal{F}$. Let $U' \in \mathcal{F}_{d_a(U)}$ be given with

$$\prod_{i=j+2}^{\infty} |U'|_i = c \leq 1.$$

First, assume $c = 1$. Since $d_a(U')^2 < d_a(U)^2$ and $|U'|_i = 1$ for all integers $i \geq j+2$, it follows that

$$\sum_{i=1}^{j+1} a_i^2 |U'|_i^2 + \sum_{i=j+2}^{\infty} a_i^2 = d_a(U')^2 < d_a(U)^2 = 2a_{j+1}^2 + \sum_{i=j+2}^{\infty} a_i^2$$

or equivalently that

$$\sum_{i=1}^{j+1} a_i^2 |U'|_i^2 < 2a_{j+1}^2.$$

If we set $G = U$, $m = j+1$, and $D = \sqrt{2}\,a_{j+1}$ in Lemma 5.17, we conclude that our assertion is true for $c = 1$.

Now, let $c < 1$. By Lemma 5.17, we immediately have

$$\mathrm{vol}(U) = \prod_{i=1}^{j+1} \frac{\sqrt{2}\,a_{j+1}}{a_i\sqrt{j+1}} \quad \text{and} \quad \mathrm{vol}(U') < c\prod_{i=1}^{j+1} \frac{d_a(U')}{a_i\sqrt{j+1}},$$

since

$$\sum_{i=1}^{j+1} a_i^2 |U'|_i^2 < \sum_{i=1}^{\infty} a_i^2 |U'|_i^2 = d_a(U')^2.$$

Therefore, we obtain

$$\begin{aligned} \frac{\mathrm{vol}(U')}{\mathrm{vol}(U)} &< c \prod_{i=1}^{j+1} \frac{d_a(U')}{a_i\sqrt{j+1}} \frac{a_i\sqrt{j+1}}{\sqrt{2}\, a_{j+1}} \\ &< c \prod_{i=1}^{j+1} \frac{d_a(U)}{\sqrt{2}\, a_{j+1}} \qquad \left(\text{because of } U' \in \mathcal{F}_{d_a(U)}\right) \\ &= c \prod_{i=1}^{j+1} \frac{\sqrt{2a_{j+1}^2 + a_{j+2}^2 + a_{j+3}^2 + \cdots}}{\sqrt{2}\, a_{j+1}} \\ &= c\left(1 + \frac{a_{j+2}^2 + a_{j+3}^2 + a_{j+4}^2 + \cdots}{2a_{j+1}^2}\right)^{(j+1)/2} \\ &< c\left(1 + \frac{a_{j+2}^2}{a_{j+1}^2}\right)^{(j+1)/2}, \end{aligned} \tag{5.29}$$

where the last strict inequality sign holds because by condition (5.16) we obtain, with a small $\varepsilon > 0$,

$$a_{j+3}^2 < \frac{\varepsilon}{j+2} a_{j+2}^2, \quad a_{j+4}^2 < \frac{\varepsilon}{j+3} a_{j+3}^2, \quad \cdots$$

and hence

$$\begin{aligned} & a_{j+3}^2 + a_{j+4}^2 + \cdots \\ & \quad < \frac{\varepsilon}{j+2} a_{j+2}^2 + \frac{\varepsilon}{j+3}\frac{\varepsilon}{j+2} a_{j+2}^2 + \frac{\varepsilon}{j+4}\frac{\varepsilon}{j+3}\frac{\varepsilon}{j+2} a_{j+2}^2 + \cdots \\ & \quad < \frac{\varepsilon}{j+2} a_{j+2}^2 \left(1 + \frac{\varepsilon}{j+2} + \left(\frac{\varepsilon}{j+2}\right)^2 + \cdots\right) \\ & \quad < a_{j+2}^2. \end{aligned}$$

It follows from (5.16) and (5.29) that

$$\frac{\mathrm{vol}(U')}{\mathrm{vol}(U)} < c\left((1+\xi_j)^{1/\xi_j}\right)^{(j+1)\xi_j/2} < 2c < 2,$$

where $\xi_j = \frac{a_{j+2}^2}{a_{j+1}^2}$ and thus $\xi_j = o\big(\frac{1}{j}\big)$ by (5.16). Consequently, it holds that

$$2\mathrm{vol}(U) > \mathrm{vol}(U'),$$

as required. □

The following lemma gives upper and lower bounds for $\mu^*(I^\omega)$.

Lemma 5.19. *If the monotone decreasing sequence $a = \{a_i\}_{i\in\mathbb{N}}$ of positive real numbers satisfies* (5.16)*, then*

$$\frac{1}{4} \le \mu^*(I^\omega) \le 1.$$

Proof. Let $\delta_k > 0$ and $\varepsilon > 0$ be sufficiently small and let $\{U_i\}_{i\in\mathbb{N}} \subset \mathcal{U}_{\delta_k}$ be a covering of I^ω satisfying

$$\sum_{i=1}^{\infty} \tau(U_i) \le \mu_k^*(I^\omega) + \varepsilon. \tag{5.30}$$

By (5.17) and (5.19), there exists, for every $i \in \mathbb{N}$, a sufficiently large $j(i) \in \mathbb{N}$ such that

$$d_a(U_i)^2 = \sum_{n=j(i)+1}^{\infty} a_n^2 + a_{j(i)+1}^2 = \delta_{j(i)}^2.$$

Now, let $U \subset I^\omega$ be a basic cylinder satisfying

$$|U|_n = \begin{cases} \dfrac{\sqrt{2}\, a_{j(i)+1}}{a_n\sqrt{j(i)+1}} & (\text{for } n \in \{1, 2, \ldots, j(i)+1\}), \\ 1 & (\text{for } n > j(i)+1). \end{cases} \tag{5.31}$$

Then $U \in \mathcal{U}$ and

$$\begin{aligned} d_a(U)^2 &= \sum_{n=1}^{\infty} a_n^2 |U|_n^2 \\ &= \sum_{n=1}^{j(i)+1} a_n^2 \frac{2a_{j(i)+1}^2}{a_n^2(j(i)+1)} + \sum_{n=j(i)+2}^{\infty} a_n^2 \\ &= 2a_{j(i)+1}^2 + \sum_{n=j(i)+2}^{\infty} a_n^2 \\ &= d_a(U_i)^2. \end{aligned}$$

In view of (5.31), for each $n \in \{1, 2, \ldots, j(i) + 1\}$, the nth edge of I^ω can be covered with at most $1 + \frac{a_n\sqrt{j(i)+1}}{\sqrt{2}\,a_{j(i)+1}}$ intervals of length $|U|_n$. Hence, by (5.20) and (5.31), we obtain

$$
\begin{aligned}
\tau(U_i) &\geq \left(\prod_{n=1}^{j(i)+1}\left(1 + \frac{a_n\sqrt{j(i)+1}}{\sqrt{2}\,a_{j(i)+1}}\right)\right)^{-1} \\
&= \prod_{n=1}^{j(i)+1} \frac{\sqrt{2}\,a_{j(i)+1}}{\sqrt{2}\,a_{j(i)+1} + a_n\sqrt{j(i)+1}}.
\end{aligned}
\tag{5.32}
$$

It now follows from (5.16), (5.30), and (5.32) that

$$
\begin{aligned}
\mu_k^*(I^\omega) &\geq \sum_{i=1}^{\infty} \tau(U_i) - \varepsilon \geq \sum_{i=1}^{\infty}\prod_{n=1}^{j(i)+1} \frac{\sqrt{2}\,a_{j(i)+1}}{\sqrt{2}\,a_{j(i)+1} + a_n\sqrt{j(i)+1}} - \varepsilon \\
&\geq \sum_{i=1}^{\infty}\left(\prod_{n=1}^{j(i)+1} \frac{\sqrt{2}\,a_{j(i)+1}}{a_n\sqrt{j(i)+1}} \cdot \left(1 - \frac{\sqrt{2}\,a_{j(i)+1}}{a_n\sqrt{j(i)+1}}\right)\right) - \varepsilon \\
&\geq \sum_{i=1}^{\infty}\left(\prod_{n=1}^{j(i)+1} \frac{\sqrt{2}\,a_{j(i)+1}}{a_n\sqrt{j(i)+1}}\right)\left(1 - \frac{\sqrt{2}\,a_{j(i)+1}}{a_{j(i)}\sqrt{j(i)+1}}\right)^{j(i)} \cdot \\
&\qquad \cdot \left(1 - \frac{\sqrt{2}}{\sqrt{j(i)+1}}\right) - \varepsilon \\
&\geq \frac{1}{2}\sum_{i=1}^{\infty}\prod_{n=1}^{j(i)+1} \frac{\sqrt{2}\,a_{j(i)+1}}{a_n\sqrt{j(i)+1}} - \varepsilon,
\end{aligned}
\tag{5.33}
$$

since it follows from (5.16) that

$$
\left(1 - \frac{\sqrt{2}\,a_{j+1}}{a_j\sqrt{j+1}}\right)^j = \left(\left(1 - \frac{\sqrt{2}\,a_{j+1}}{a_j\sqrt{j+1}}\right)^{-\frac{a_j\sqrt{j+1}}{\sqrt{2}\,a_{j+1}}}\right)^{-\frac{\sqrt{2}\,j a_{j+1}}{a_j\sqrt{j+1}}} \geq \frac{1}{2}
$$

for any sufficiently large j. Since $\{U_i\}_{i\in\mathbb{N}}$ is a covering of I^ω, by (5.33) and Lemma 5.18, we get

$$
\mu_k^*(I^\omega) \geq \frac{1}{2}\sum_{i=1}^{\infty}\frac{1}{2}\sup\left\{\mathrm{vol}(U') : U' \in \mathcal{G}_{d_a(U_i)}\right\} - \varepsilon \geq \frac{1}{4} - \varepsilon
$$

and

$$
\mu^*(I^\omega) = \sup\left\{\mu_k^*(I^\omega) : k \in \mathbb{N}\right\} \geq \frac{1}{4} - \varepsilon.
$$

Since $\varepsilon > 0$ can be chosen arbitrarily small, this implies

$$\mu^*(I^\omega) \geq \frac{1}{4}.$$

On the other hand, for each sufficiently large $k \in \mathbb{N}$, there exists a covering $\{F_i\}_{i\in\{1,2,\ldots,n\}} \subset \mathcal{F}_{\delta_{k-1}}$ of I^ω such that $d_a(F_i) = \delta_k$ for all $i \in \{1, 2, \ldots, n\}$ and $n = N'(I^\omega, \delta_{k-1})$. Since a is a monotone decreasing sequence of positive real numbers, $\delta_k^2 = \sum_{j=k+1}^{\infty} a_j^2 + a_{k+1}^2 \leq \delta_{k-1}^2 = \sum_{j=k}^{\infty} a_j^2 + a_k^2$, $\mathcal{U}_{\delta_k} \subset \mathcal{U}_{\delta_{k-1}}$, and since $N'(I^\omega, \delta_k) \geq N'(I^\omega, \delta_{k-1})$, we get $N'(I^\omega, d_a(F_i)) \geq n$, $\tau(F_i) = \frac{1}{N'(I^\omega, d_a(F_i))} \leq \frac{1}{n}$, $\sum_{i=1}^{n} \tau(F_i) \leq \sum_{i=1}^{n} \frac{1}{n} = 1$, and thus, $\mu^*(I^\omega) \leq 1$, as desired. □

In the following lemma, we show that the restriction of μ^* to the σ-algebra of all μ^*-measurable subsets of I^ω, denoted by μ', is d_a-invariant. We remark that each Borel set in I^ω is μ^*-measurable (see Remark 5.16 (ii)).

We note that the paper [11] has been improved by replacing [11, Lemma 2.10] and [11, Lemma 3.1] with Theorem 4.26 and the following lemma, respectively.

Lemma 5.20. *If the monotone decreasing sequence $a = \{a_i\}_{i\in\mathbb{N}}$ of positive real numbers satisfies the condition* (5.16)*, then μ' is d_a-invariant.*

Proof. Let A and B be Borel subsets of I^ω that are d_a-isometric to each other via a (fixed) surjective d_a-isometry $f : A \to B$. Without loss of generality, we assume that A is non-degenerate. Then, by Theorem 4.26, $\mathrm{GS}^2(A, p)$ includes every cylinder from $\mathcal{G}$, where p is a (fixed) element of A. (When A is degenerate, we can revise Theorem 4.26 by replacing $\mathbb{N}$ with $\Lambda(A)$. Then, $\mathrm{GS}^2(A, p) - p$ is a proper subspace of M_a.) If $F_2 : \mathrm{GS}^2(A, p) \to M_a$ is given as in Definition 4.14, then F_2 is a d_a-isometry and it extends the surjective d_a-isometry $f : A \to B$ by Theorem 4.27.

If $\{U_i\}_{i\in\mathbb{N}}$ is a covering of A with each U_i from $\mathcal{G}$, then it follows from (5.18) that there exist a B_i from $\mathcal{F}$ and a surjective d_a-isometry $f_i : B_i \to U_i$ for each i. In particular, the cylinder $U_i = f_i(B_i)$ from $\mathcal{G}$ is included in $\mathrm{GS}^2(A, p)$, the domain of F_2, by the argument in the first part (or by Theorem 4.26). Furthermore, we know that $\{F_2(U_i)\}_{i\in\mathbb{N}} = \{(F_2 \circ f_i)(B_i)\}_{i\in\mathbb{N}}$ is a covering of B with $d_a(U_i) = d_a(F_2(U_i))$.

Due to the definitions of μ' and $\mathcal{U}_\delta$ and by the previous argument, we have

$$\begin{aligned}
\mu'(A) &= \sup_{k\in\mathbb{N}} \mu_k^*(A) \\
&= \sup_{k\in\mathbb{N}} \inf \left\{ \sum_{i=1}^{\infty} \tau(U_i) : \{U_i\}_{i\in\mathbb{N}} \subset \mathcal{U}_{\delta_k} \text{ is a covering of } A \right\} \\
&= \sup_{k\in\mathbb{N}} \inf \left\{ \sum_{i=1}^{\infty} \tau\big(F_2(U_i)\big) : \{U_i\}_{i\in\mathbb{N}} \subset \mathcal{U}_{\delta_k} \text{ is a covering of } A \right\} \\
&= \sup_{k\in\mathbb{N}} \inf \left\{ \sum_{i=1}^{\infty} \tau\big(F_2(U_i)\big) : \{F_2(U_i)\}_{i\in\mathbb{N}} \subset \mathcal{U}_{\delta_k} \text{ is a covering of } B \right\} \\
&\geq \sup_{k\in\mathbb{N}} \inf \left\{ \sum_{i=1}^{\infty} \tau(V_i) : \{V_i\}_{i\in\mathbb{N}} \subset \mathcal{U}_{\delta_k} \text{ is a covering of } B \right\} \\
&= \mu'(B),
\end{aligned}$$

since some coverings of B may not be represented as images of coverings of A under F_2. Analogously, we can easily prove the opposite inequality. □

For all Borel sets $C \subset I^\omega$, let

$$\nu(C) = \frac{\mu'(C)}{\mu'(I^\omega)}.$$

Clearly, in view of Lemma 5.20, the measure ν is d_a-invariant with $\nu(I^\omega) = 1$.

The proof of the following lemma is comparable to that of [17, Lemma 1].

Lemma 5.21. *Assume that the monotone decreasing sequence $a = \{a_i\}_{i\in\mathbb{N}}$ of positive real numbers satisfies the condition* (5.16). *Then, the measure ν coincides with the standard product probability measure π on the Borel subsets of I^ω.*

Proof. Given any positive integers m and n, let $\mathcal{Z}_{mn}$ be the collection of the non-degenerate basic cylinders C (in I^ω) defined by

$$C = \prod_{i=1}^{\infty} C_i,$$

where

$$C_i = \begin{cases} \left[\frac{j}{m}, \frac{j+1}{m}\right] \text{ for some } j \in \{0, 1, \ldots, m-1\} & (\text{for } i \leq n), \\ [0, 1] & (\text{for } i > n). \end{cases}$$

The translation invariance of d_a implies that of ν. Thus we have

$$\nu(C_1) = \nu(C_2) \tag{5.34}$$

for all $C_1, C_2 \in \mathcal{Z}_{mn}$.

Let C be an arbitrary basic cylinder in $\mathcal{Z}_{mn}$. In general, we let

$$C = \big\{(x_1, x_2, \ldots) \in I^\omega : b_i \leq x_i \leq e_i \text{ for all } i \in \mathbb{N}\big\},$$

where $0 \leq b_i < e_i \leq 1$ for $i \in \{1, 2, \ldots, n\}$ and $b_i = 0$ and $e_i = 1$ for $i > n$. Let ∂C denote the boundary of C. Then, there are at most countably many sets $H_1, H_2, \ldots$ of the forms

$$H_{2i-1} = \big\{(x_1, x_2, \ldots) \in I^\omega : x_i = b_i\big\}$$

and

$$H_{2i} = \big\{(x_1, x_2, \ldots) \in I^\omega : x_i = e_i\big\}$$

satisfying

$$\partial C \subset \bigcup_{k=1}^{\infty} H_k. \tag{5.35}$$

It is obvious that there are infinitely many disjoint translates of H_k in I^ω for every $k \in \mathbb{N}$. Since $\frac{1}{4} \leq \mu^*(I^\omega) \leq 1$ by Lemma 5.19, we obtain

$$\mu^*(H_k) = 0$$

for each $k \in \mathbb{N}$. Due to Theorem 3.6, H_k is μ^*-measurable. Hence, $\mu'(H_k) = 0$ and

$$\nu(H_k) = \frac{\mu'(H_k)}{\mu'(I^\omega)} = 0. \tag{5.36}$$

It follows from (5.35) and (5.36) that

$$\nu(\partial C) = 0 \tag{5.37}$$

for all $C \in \mathcal{Z}_{mn}$.

In addition, on account of Remark 5.16 (ii), we conclude that all Borel subsets of I^ω are μ^*-measurable. Hence, by (5.37), we have

$$\nu(C) = \nu\big(C^\circ\big)$$

for each $C \in \mathcal{Z}_{mn}$. Therefore, we get

$$\nu\left(\bigcup_{C\in\mathcal{Z}_{mn}} C^\circ\right) \leq \nu(I^\omega) \leq \nu\left(\bigcup_{C\in\mathcal{Z}_{mn}} C\right) \leq \nu\left(\bigcup_{C\in\mathcal{Z}_{mn}} C^\circ\right),$$

and by (5.34), we have

$$\nu(C) = \nu(C^\circ) = \frac{1}{m^n} = \pi(C) \tag{5.38}$$

for all $C \in \mathcal{Z}_{mn}$.

Finally, given an $n \in \mathbb{N}$, let J_n be an arbitrary non-degenerate basic cylinder defined by

$$J_n = \prod_{i=1}^{\infty} [p_{1i}, p_{2i}],$$

where $0 \leq p_{1i} < p_{2i} \leq 1$ for any $i \in \{1, 2, \ldots, n\}$ and $p_{1i} = 0$, $p_{2i} = 1$ for any $i > n$. For all positive integers m, we define the finite collection $\mathcal{C}_{mn}$ of basic cylinders by

$$\mathcal{C}_{mn} = \{C \in \mathcal{Z}_{mn} : C \cap J_n \neq \emptyset\}.$$

Then, every non-degenerate basic cylinder J_n can be expressed as

$$J_n = \bigcap_{m=1}^{\infty} \bigcup_{C\in\mathcal{C}_{mn}} C,$$

where the finite union $\bigcup_{C\in\mathcal{C}_{mn}} C$ of basic cylinders is closed.

Since the collection of all non-degenerate basic cylinders J_n for all $n \in \mathbb{N}$, together with the empty set, generates the Borel σ-algebra over I^ω, we conclude that the basic cylinders from $\bigcup_{m,n=1}^{\infty} \mathcal{Z}_{mn}$, together with the empty set, generate the Borel σ-algebra over I^ω. Therefore, (5.38) implies that ν coincides with π on the Borel subsets of I^ω. □

As we already mentioned in the paragraph following Lemma 5.20, the measure ν is d_a-invariant. Using Lemma 5.21, we obtain the following result:

Theorem 5.22. *For any monotone decreasing sequence $a = \{a_i\}_{i\in\mathbb{N}}$ of positive real numbers satisfying* (5.16), *the standard product probability measure π on I^ω is d_a-invariant.*

Chapter 6

Ulam's Conjecture

The conjecture of Ulam states that the standard product probability measure π on the Hilbert cube I^ω is invariant under the induced metric d_a when the sequence $a = \{a_i\}_{i \in \mathbb{N}}$ of positive numbers satisfies condition (4.1). This conjecture was proved in [6] when E_1 is a non-degenerate subset of M_a. In this chapter, we will completely prove Ulam's conjecture to be true by considering both non-degenerate as well as degenerate cases. More precisely, under the assumption that the axiom of choice is accepted and the sequence $a = \{a_i\}_{i \in \mathbb{N}}$ of positive real numbers satisfies the condition $\sum_{i=1}^{\infty} a_i^2 < \infty$, we prove that $\pi(E_1) = \pi(E_2)$ for all Borel subsets E_1 and E_2 of I^ω which are d_a-isometric to each other, where π is the standard product probability measure on I^ω.

6.1 Basic Definitions

As we did in Sects. 4.1 and 5.2, we denote by $\mathbb{R}^\omega$ the infinite-dimensional real vector space defined as

$$\mathbb{R}^\omega = \big\{ (x_1, x_2, \ldots) : x_i \in \mathbb{R} \text{ for all } i \in \mathbb{N} \big\}.$$

In addition, $(\mathbb{R}^\omega, \mathcal{T})$ denotes the product space $\prod_{i=1}^{\infty} \mathbb{R}$, where $(\mathbb{R}, \mathcal{T}_{\mathbb{R}})$ is the usual topological space.

Let $I = [0, 1]$ be the unit closed interval, $I^\omega = \prod_{i=1}^{\infty} I$ the *Hilbert cube*, and let π be the *standard product probability measure* on I^ω. We denote by $(I^\omega, \mathcal{T}_\omega)$ the (topological) subspace of $(\mathbb{R}^\omega, \mathcal{T})$. Then, $\mathcal{T}_\omega$ is the relative topology for I^ω induced by $\mathcal{T}$.

S.-M. Jung, *Ulam's Conjecture on Invariance of Measure in the Hilbert Cube*, Frontiers in Mathematics, https://doi.org/10.1007/978-3-031-30886-4_6

As in Sect. 4.1, let $a = \{a_i\}_{i \in \mathbb{N}}$ be a sequence of positive real numbers that satisfies the condition

$$\sum_{i=1}^{\infty} a_i^2 < \infty.$$

Using this sequence a, we define the metric on I^ω by

$$d_a(x, y) = \left(\sum_{i=1}^{\infty} a_i^2 (x_i - y_i)^2 \right)^{1/2}$$

for all $x = (x_1, x_2, \ldots) \in I^\omega$ and $y = (y_1, y_2, \ldots) \in I^\omega$.

As in Sect. 4.1, we define the real Hilbert space M_a by

$$M_a = \left\{ (x_1, x_2, \ldots) \in \mathbb{R}^\omega : \sum_{i=1}^{\infty} a_i^2 x_i^2 < \infty \right\}$$

and we define an inner product $\langle \cdot, \cdot \rangle_a$ on M_a by

$$\langle x, y \rangle_a = \sum_{i=1}^{\infty} a_i^2 x_i y_i$$

for all $x = (x_1, x_2, \ldots)$ and $y = (y_1, y_2, \ldots)$ of M_a. Then this inner product induces the norm

$$\|x\|_a = \sqrt{\langle x, x \rangle_a}$$

for all $x \in M_a$.

In view of definition (4.2), the metric d_a on I^ω can be extended to the metric on M_a, i.e.,

$$d_a(x, y) = \sqrt{\langle x - y, x - y \rangle_a}$$

for all $x, y \in M_a$.

6.2 Cylinders

For each positive integer n, let $\mathcal{B}_n$ be the set of all basic cylinders $J = \prod_{i=1}^{\infty} J_i$ defined by Definition 4.20 for some disjoint finite subsets Λ_1, Λ_2, Λ_3, Λ_4 of $\{1, 2, \ldots, n\}$ and $0 < p_{1i} < p_{2i} < 1$ for $i \in \Lambda_1 \cup \Lambda_2 \cup \Lambda_3$ and $0 \leq p_{1i} \leq 1$ for $i \in \Lambda_4$. This definition of basic cylinders is a slight modification of Definition 4.20, but the two definitions are essentially the same. We note that $\Lambda_1 \cup \Lambda_2 \cup \Lambda_3 \cup \Lambda_4 \subset \{1, 2, \ldots, n\}$ and at most n edges of each basic cylinder of $\mathcal{B}_n$ have a Euclidean length of less than 1, and all remaining edges have a Euclidean length of 1.

On the other hand, in some cases the set $\Lambda_1 \cup \Lambda_2 \cup \Lambda_3 \cup \Lambda_4$ can be infinite. In this case, the corresponding interval will be called the *extraordinary basic cylinder*.

Example 6.1. Let $J = \{0\} \times \prod_{i=2}^{\infty} [0, 1]$ be a degenerate basic cylinder and let $K = \prod_{i=1}^{\infty} \big[0, \frac{a_{i+1}}{a_i}\big]$ be an infinite-dimensional interval. If we define a function $f : J \to K$ by

$$f\left(\frac{1}{a_{i+1}} e_{i+1}\right) = \frac{1}{a_i} e_i,$$

for all $i \in \mathbb{N}$, and by

$$f(x) = \sum_{i=1}^{\infty} a_{i+1} x_{i+1} \frac{1}{a_i} e_i$$

for all $x = \sum_{i=1}^{\infty} x_i e_i = \sum_{i=1}^{\infty} a_i x_i \frac{1}{a_i} e_i \in J$, then

$$\begin{aligned} \|f(x) - f(y)\|_a^2 &= \left\langle \sum_{i=1}^{\infty} a_{i+1}(x_{i+1} - y_{i+1}) \frac{1}{a_i} e_i, \sum_{j=1}^{\infty} a_{j+1}(x_{j+1} - y_{j+1}) \frac{1}{a_j} e_j \right\rangle_a \\ &= \sum_{i=2}^{\infty} a_i^2 (x_i - y_i)^2 = \|x - y\|_a^2 \end{aligned}$$

for all $x = \sum_{i=1}^{\infty} x_i e_i \in J$ and $y = \sum_{i=1}^{\infty} y_i e_i \in J$. That is, f is a d_a-isometry from J onto K. Thus, $K = \prod_{i=1}^{\infty} \big[0, \frac{a_{i+1}}{a_i}\big]$ is a cylinder as an isometric image of a degenerate basic cylinder J. We note that K is an extraordinary basic cylinder.

For $i \in \{1, 2, \ldots, n\}$, each J_i is a compact subinterval of $[0, 1]$ with respect to the relative topology for $[0, 1]$. Thus, by Theorems 1.20, 1.29, and 1.36, the infinite-dimensional interval

$$\prod_{1 \le j < i} [0, 1] \times J_i \times \prod_{j > i} [0, 1]$$

is a compact subset of the Hilbert cube I^ω. Since I^ω is a Hausdorff space, Theorem 1.43 implies that the infinite-dimensional interval above is a closed subset of the Hilbert cube I^ω for each $i \in \{1, 2, \ldots, n\}$. In particular, when $i = 1$, we read the last expression as

$$\prod_{1 \le j < 1} [0,1] \times J_1 \times \prod_{j > 1} [0,1] = J_1 \times \prod_{j > 1} [0,1].$$

Since every basic cylinder $J \in \mathcal{B}_n$ is expressed as

$$J = \prod_{i=1}^{\infty} J_i = \bigcap_{i=1}^{n} \left(\prod_{1 \le j < i} [0,1] \times J_i \times \prod_{j > i} [0,1] \right),$$

we see that each $J \in \mathcal{B}_n$ is a closed subset of I^ω as the intersection of closed sets. Since I^ω is a closed subset of M_a by Remark 4.5 (iv), we use Theorem 1.46 to conclude that J is a closed subset of M_a.

Now, let us define

$$\mathcal{B} = \{\emptyset\} \cup \bigcup_{n=1}^{\infty} \mathcal{B}_n \quad \text{and} \quad \mathcal{B}_\delta = \{J \in \mathcal{B} : d_a(J) < \delta\}$$

for any $0 < \delta < 1$, where the *diameter* of J is defined as $d_a(J) = \sup\{d_a(x, y) : x, y \in J\}$. Here we need to distinguish the difference between $\mathcal{B}_n$ for the positive integer n and $\mathcal{B}_\delta$ for the real number $0 < \delta < 1$. We remark that every basic cylinder in $\mathcal{B}$ is a closed subset of M_a. (Also, each extraordinary basic cylinder is likewise a closed subset of M_a.)

We denote by $\mathcal{C}$ the set of every subset K of M_a, for which there exists a basic cylinder $J \in \mathcal{B}$ and a surjective d_a-isometry $f : J \to K$, and we define $\mathcal{C}_\delta = \{K \in \mathcal{C} : d_a(K) < \delta\}$ for any $\delta > 0$. We notice that $\mathcal{B}_\delta \subset \mathcal{C}_\delta$ for every $0 < \delta < 1$. We note that the family $\mathcal{B}$ includes not only non-degenerate basic cylinders but also degenerate ones.

Assume that a basic cylinder J and a cylinder K are given such that J is d_a-isometric to K through a surjective d_a-isometry $f : J \to K$. Since J is a compact subset of M_a as a closed subset of a compact set I^ω, K is also a compact subset of M_a as the continuous image of a compact set J (see Theorem 1.23). Moreover, K is a closed subset of M_a as a compact subset of the Hausdorff space M_a (see Theorem 1.43). As we did in Remark 5.15, we come to an important consequence.

Remark 6.2. Every cylinder $K \in \mathcal{C}$ is closed in M_a.

Let J be a basic cylinder that is d_a-isometric to a cylinder K via a surjective d_a-isometry $f : J \to K$. Assume that $p = (p_1, p_2, \ldots)$ is the lower left corner of J, q is an element of K with $q = f(p)$, $\mathrm{GS}(J, p)$ is the first-order generalized span of J with respect to p, and that $F : \mathrm{GS}(J, p) \to M_a$ is the extension of f given in Definition 4.11. If $x = (x_1, x_2, \ldots) \in \mathrm{GS}(J, p)$, then it follows from

Theorem 4.24 that $x - p = \sum_{j=1}^{\infty}(x_j - p_j)e_j = \sum_{j\in\Lambda(J)}(x_j - p_j)e_j \in M_a$. By Theorem 4.31, we get

$$\begin{aligned}
&(T_{-q} \circ F \circ T_p)(x-p) \\
&= \sum_{i\in\Lambda(J)} \left\langle x-p, \frac{1}{a_i}e_i \right\rangle_a \frac{1}{a_i}(T_{-q} \circ F \circ T_p)(e_i) \\
&= \sum_{i\in\Lambda(J)} a_i(x_i - p_i)\frac{1}{a_i}(T_{-q} \circ F \circ T_p)(e_i).
\end{aligned} \tag{6.1}$$

For any $y = (y_1, y_2, \ldots) \in \mathrm{GS}(J, p)$, it follows from Theorem 4.24 that

$$\begin{aligned}
y &= p + (y-p) \\
&= p + \sum_{i=1}^{\infty}(y_i - p_i)e_i \\
&= p + \sum_{i\in\Lambda(J)} a_i(y_i - p_i)\frac{1}{a_i}e_i \\
&\in \mathrm{GS}(J, p)
\end{aligned}$$

and it further follows from (6.1) that

$$\begin{aligned}
F(y) &= q + (T_{-q} \circ F \circ T_p)(y-p) \\
&= q + \sum_{i\in\Lambda(J)} a_i(y_i - p_i)\frac{1}{a_i}(T_{-q} \circ F \circ T_p)(e_i) \\
&\in \mathrm{GS}(K, q),
\end{aligned}$$

since $\mathrm{GS}(K, q) = F(\mathrm{GS}(J, p))$ by Theorem 4.32. Moreover, since the sequences $\{\frac{1}{a_i}e_i\}_{i\in\Lambda(J)}$ and $\{\frac{1}{a_i}(T_{-q}\circ F\circ T_p)(e_i)\}_{i\in\Lambda(J)}$ are both orthonormal, the following definition may be useful.

Definition 6.3. Every interval in $\mathcal{B}$ will be called a *basic cylinder* and each element of $\mathcal{C}$ a *cylinder*. Assume that J is a basic cylinder. If $\Lambda(J) = \mathbb{N}$, then J will be called *non-degenerate*. Otherwise, it will be called *degenerate*.

Remark 6.4. The term "non-degenerate" or "degenerate" defined in relation to basic cylinders is similar to, but not identical to, the term "non-degenerate" or "degenerate" for the general sets E defined in Definition 4.8. We notice that the terms "non-degenerate" and "degenerate" are used for convenience only and are not exact mathematical terms.

6.3 Elementary Volumes

Assume that both J_1 and J_2 are two distinct basic cylinders which are d_a-isometric to the same cylinder K via the surjective d_a-isometries $f_1 : J_1 \to K$ and $f_2 : J_2 \to K$, respectively. Moreover, assume that u is the lower left corner and x is the vertex of J_1 diagonally opposite to u, i.e., x is the upper right corner of J_1. Analogously, let v be the lower left corner and y the vertex of J_2 diagonally opposite to v and $f_1(u) = f_2(v) =: w \in K$. Furthermore, assume that $F_{(1)} : \mathrm{GS}(J_1, u) \to M_a$ and $F_{(2)} : \mathrm{GS}(J_2, v) \to M_a$ are d_a-isometries given in Definition 4.11 and that they are extensions of f_1 and f_2, respectively.

Then, by (6.1), we have

$$\begin{aligned}(T_{-w} \circ F_{(1)} \circ T_u)(x-u) &= (T_{-w} \circ F_{(1)} \circ T_u)\left(\sum_{i\in\Lambda(J_1)} (x_i - u_i)e_i\right)\\ &= \sum_{i\in\Lambda(J_1)} (x_i - u_i)(T_{-w} \circ F_{(1)} \circ T_u)(e_i),\\ (T_{-w} \circ F_{(2)} \circ T_v)(y-v) &= (T_{-w} \circ F_{(2)} \circ T_v)\left(\sum_{i\in\Lambda(J_2)} (y_i - v_i)e_i\right)\\ &= \sum_{i\in\Lambda(J_2)} (y_i - v_i)(T_{-w} \circ F_{(2)} \circ T_v)(e_i).\end{aligned} \tag{6.2}$$

Further, the right hand side of the first equality expresses the vector $f_1(x) - w$, since

$$f_1(x) - w = F_{(1)}(x) - w = (T_{-w} \circ F_{(1)} \circ T_u)(x-u).$$

Similarly, the right hand side of the second equality in (6.2) expresses the vector $f_2(y) - w$.

According to (4.26), Theorem 4.31 and (6.2), the coordinates $\left\langle x-u, \frac{1}{a_i}e_i\right\rangle_a$ remain unchanged under the action of the d_a-isometry $T_{-w} \circ F_{(1)} \circ T_u$. Moreover, the points w and $f_1(x)$ are the diagonally opposite vertices of the cylinder K. The same is true for w and $f_2(y)$. Thus, we can conclude that $f_1(x) - w = f_2(y) - w$.

Since $(T_{-w} \circ F_{(1)} \circ T_u)((x_i - u_i)e_i) = (x_i - u_i)(T_{-w} \circ F_{(1)} \circ T_u)(e_i)$ for each $i \in \Lambda(J_1)$, $F_{(1)}$ maps each edge of basic cylinder J_1 onto the edge of K. Conversely, every edge of the cylinder K is an image of the edge of J_1 under the d_a-isometry $F_{(1)}$. The same case is also for $F_{(2)}$ and J_2. Therefore, we conclude that there exists a permutation $\sigma : \Lambda(J_1) \to \Lambda(J_2)$ that satisfies

$$(x_i - u_i)(T_{-w} \circ F_{(1)} \circ T_u)(e_i) = (y_{\sigma(i)} - v_{\sigma(i)})(T_{-w} \circ F_{(2)} \circ T_v)(e_{\sigma(i)})$$

for any $i \in \Lambda(J_1)$.

According to (4.10), it is obvious that both $\left\{\frac{1}{a_i}(T_{-w} \circ F_{(1)} \circ T_u)(e_i)\right\}$ and $\left\{\frac{1}{a_i}(T_{-w} \circ F_{(2)} \circ T_v)(e_i)\right\}$ are orthonormal sequences. Hence, we get

$$a_i|x_i - u_i| = a_{\sigma(i)}|y_{\sigma(i)} - v_{\sigma(i)}| \tag{6.3}$$

for all $i \in \Lambda(J_1)$. Since J_1 and J_2 are basic cylinders, due to the structural property of basic cylinders, we see that there is an $\ell_0 \in \mathbb{N}$ such that $|x_i - u_i| = |y_i - v_i| = 1$ for all $i > \ell_0$. Thus, it follows from (6.3) that there exists an $m_0 \in \mathbb{N}$ ($m_0 \geq \ell_0$) that satisfies $a_i = a_{\sigma(i)}$ for each $i > m_0$.

Consequently, when J_1 is non-degenerate, we use (6.3) to show that the basic cylinders J_1 and J_2 have the same *elementary volume*:

$$\begin{aligned}
\mathrm{vol}(J_1) &= \prod_{i=1}^{\infty} |x_i - u_i| = \prod_{i=1}^{\infty} \frac{a_{\sigma(i)}}{a_i} |y_{\sigma(i)} - v_{\sigma(i)}| \\
&= \prod_{i=1}^{m_0} \frac{a_{\sigma(i)}}{a_i} |y_{\sigma(i)} - v_{\sigma(i)}| \times \prod_{i=m_0+1}^{\infty} |y_{\sigma(i)} - v_{\sigma(i)}| \\
&= \prod_{i=1}^{m_0} \frac{a_{\sigma(i)}}{a_i} \times \prod_{i=1}^{\infty} |y_{\sigma(i)} - v_{\sigma(i)}| \\
&= \prod_{i=1}^{m_0} \frac{a_{\sigma(i)}}{a_i} \times \prod_{i=1}^{\infty} |y_i - v_i| \\
&= \prod_{i=1}^{\infty} |y_i - v_i| \\
&= \mathrm{vol}(J_2).
\end{aligned}$$

Hence, it is reasonable to define the *volume* of the cylinder K as the elementary volume of one of the basic cylinders which are d_a-isometric to K, i.e.,

$$\mathrm{vol}(K) = \mathrm{vol}(J_1) = \mathrm{vol}(J_2).$$

When J_1 is degenerate, we define $\mathrm{vol}(K) = \mathrm{vol}(J_1) = \mathrm{vol}(J_2) = 0$.

Remark 6.5. Let J_1 be a basic cylinder and K_1 a cylinder. Assume that J_1 and K_1 are d_a-isometric to each other via a surjective d_a-isometry $f : J_1 \to K_1$. Assume that p is an element of J_1 and q is an element of K_1 with $q = f(p)$. Comparing (4.26) and Theorem 4.31 and considering the fact that $\mathrm{GS}(J_1, p) = \mathrm{GS}^n(J_1, p)$ for any $n \in \mathbb{N}$ (see the proof of Theorem 4.34), under the action of the d_a-isometry $T_{-q} \circ F \circ T_p : \mathrm{GS}(J_1, p) - p \to M_a$, the following statements are true.

(i) The orthonormal sequence $\left\{\frac{1}{a_i}e_i\right\}_{i\in\Lambda(J_1)}$ is changed to another orthonormal sequence $\left\{\frac{1}{a_i}(T_{-q}\circ F\circ T_p)(e_i)\right\}_{i\in\Lambda(J_1)}$.

(ii) The coordinates (or Fourier coefficients) $\left\langle u-p, \frac{1}{a_i}e_i\right\rangle_a$, $i\in\Lambda(J_1)$, of each element $u-p\in \mathrm{GS}(J_1,p)-p$ remain unchanged.

According to Theorem 4.31, the d_a-isometry $T_{-q}\circ F\circ T_p$ transforms the ith coordinate $\frac{1}{a_i}e_i$ into $\frac{1}{a_i}(T_{-q}\circ F\circ T_p)(e_i)$ for every $i\in\Lambda(J_1)$. Moreover, by (4.26) and Theorem 4.31, the coordinate expression of the image of $u-p$ under the action of $T_{-q}\circ F\circ T_p$ (in the coordinate system $\left\{\frac{1}{a_i}(T_{-q}\circ F\circ T_p)(e_i)\right\}_{i\in\Lambda(J_1)}$) is the same as that of $u-p$ in the coordinate system $\left\{\frac{1}{a_i}e_i\right\}_{i\in\Lambda(J_1)}$. Therefore, $T_{-q}\circ F\circ T_p$ preserves each m-face of each basic cylinder J contained in $\mathrm{GS}(J_1,p)-p$, where $m\in\{0,1,2,\ldots\}$. More precisely,

(iii) F maps each m-face of basic cylinder J contained in $\mathrm{GS}(J_1,p)$ onto an m-face of cylinder $F(J)$, where $m\in\{0,1,2,\ldots\}$. In particular, F maps each 1-face of J onto a 1-face of $F(J)$. Consequently, the "volume" of cylinder $F(J)$ is defined as the elementary volume of the basic cylinder J, i.e., $\mathrm{vol}(F(J)) = \mathrm{vol}(J) = \prod_{i=1}^{\infty} s_i$, where s_i is the Euclidean length of the ith edge of J.

(iv) The adjacent edges of the cylinder $F(J)$ meet orthogonally, and the volume $\mathrm{vol}(F(J))$ of $F(J)$ is the infinite product of the Euclidean lengths of all edges of $F(J)$.

Based on Remark 6.5, we can define the elementary volume of basic cylinders and the volume of cylinders accurately.

Definition 6.6. (i) The *elementary volume* of a basic cylinder J is denoted by $\mathrm{vol}(J)$ and defined by

$$\mathrm{vol}(J) = \begin{cases} \prod_{i=1}^{\infty} s_i & (\text{for } \Lambda(J)=\mathbb{N}), \\ 0 & (\text{for } \Lambda(J)\neq\mathbb{N}), \end{cases}$$

where s_i is the Euclidean length of the ith edge of J.

(ii) Considering Remark 6.5 (iii), we define the *volume* of cylinder K by

$$\mathrm{vol}(K) = \mathrm{vol}(J)$$

for any $K\in\mathcal{C}$ for which there exists a basic cylinder $J\in\mathcal{B}$ and a surjective d_a-isometry $f: J\to K$.

6.4 Construction of Invariant Measure

If we set $\mathrm{vol}(\emptyset) = 0$, then the volume "vol" defined in Definition 6.6 is a *pre-measure* (see Definition 3.10). According to Theorem 3.24 and (3.14), we can apply Munroe's Method II to construct an outer measure μ^* from the pre-measure "vol" by the formula

$$\mu^*(E) = \lim_{\delta \to 0+} \mu_\delta^*(E) \tag{6.4}$$

for all subsets E of M_a, where

$$\mu_\delta^*(E) = \inf \left\{ \sum_{i=1}^{\infty} \mathrm{vol}(C_i) : E \subset \bigcup_{i=1}^{\infty} C_i \text{ where } C_i \in \mathcal{C}_\delta \text{ for all } i \in \mathbb{N} \right\}.$$

We remark that since M_a is a separable metric space, Theorem 1.12 (Lindelöf's theorem) allows us to consider only the countable coverings.

Remark 6.7. For any extraordinary basic cylinder J_{ex} with $d_a(J_{ex}) < \delta$, there exists a basic cylinder J with $J_{ex} \subset J$ and $d_a(J) < \delta$. For example, if J_{ex} is an extraordinary basic cylinder given by $J_{ex} = \prod_{i=1}^{\infty} [\alpha_i, \beta_i]$, where $0 \le \alpha_i \le \beta_i \le 1$ for all $i \in \mathbb{N}$, then we can choose a basic cylinder J as

$$J = \prod_{i=1}^{n} [\alpha_i, \beta_i] \times \prod_{i=n+1}^{\infty} [0, 1].$$

It is obvious that J_{ex} is included in J. Moreover, if $d_a(J_{ex}) < \delta$ then we can choose a sufficiently large integer n such that $d_a(J) < \delta$.

We remind us that a collection $\{C_i : i \in \mathbb{N}\}$ of sets is called a *covering* of E if $E \subset \bigcup_{i=1}^{\infty} C_i$. If, in addition, the diameter of each C_i is less than δ, then the collection $\{C_i : i \in \mathbb{N}\}$ is called a *δ-covering* of E.

In this section, we assume that each of the sets E_1 and E_2 has uncountably many elements. We note that if E_1 and E_2 have only countably many elements, then obviously $\mu^*(E_1) = 0 = \mu^*(E_2)$.

One of the most important theorems in this book is the following theorem stating that the outer measure μ^* is d_a-invariant. However, we note that this theorem has already been proved for the non-degenerate case in [6, Theorem 3.1]. Now we will completely prove this theorem by providing the proof for the degenerate case also.

Theorem 6.8. *If E_1 and E_2 are subsets of I^ω that are d_a-isometric to each other, then $\mu^*(E_1) = \mu^*(E_2)$.*

Proof. (a) We assume that E_1 and E_2 are arbitrary subsets of I^ω, each of which has uncountably many elements, and that they are d_a-isometric to each other via the surjective d_a-isometry $f : E_1 \to E_2$. Using the definition of F (Definition 4.11) and assuming that p is an element of E_1 and q is an element of E_2 with $q = f(p)$, Theorem 4.13 states that $F : \mathrm{GS}(E_1, p) \to M_a$ is a d_a-isometry which extends the surjective d_a-isometry $f : E_1 \to E_2$.

Let r be a positive real number satisfying $E_1 \subset B_r(p)$, where $B_r(p)$ denotes the open ball defined as $B_r(p) = \{y \in M_a : \|y - p\|_a < r\}$. According to Theorem 4.27, the function $F_2 : \mathrm{GS}^2(E_1, p) \to M_a$ (defined in Definition 4.14) is a d_a-isometry and it is an extension of F and so F_2 is obviously an extension of f.

(b) We consider the case where $\Lambda(\mathrm{GS}^2(E_1, p)) \neq \mathbb{N}$. In other words, we assume that the second-order generalized span $\mathrm{GS}^2(E_1, p)$ of E_1 with respect to p is degenerate, i.e., $\mathrm{GS}^2(E_1, p) \neq M_a$ (by Theorem 4.26).

The translation $T_{-p} : M_a \to M_a$ is a homeomorphism and hence, it is a closed mapping. According to Lemma 4.17 (iii), $\mathrm{GS}^2(E_1, p)$ is a closed proper subset of M_a, so is $T_{-p}(\mathrm{GS}^2(E_1, p))$. That is, $\mathrm{GS}^2(E_1, p) - p$ is a closed (proper) subspace of the real Hilbert space M_a and hence, it follows from Theorem 1.41 that $\mathrm{GS}^2(E_1, p) - p$ is itself a real Hilbert space.

$(b.1)$ In this part we assume the axiom of choice. Let $\{\frac{1}{a_i} e_i\}_{i \in \Lambda}$ be a complete orthonormal sequence in the Hilbert space $\mathrm{GS}^2(E_1, p) - p$, where Λ is a nonempty proper subset of $\mathbb{N}$. Moreover, we note that M_a can be orthogonally decomposed into

$$M_a = \big(\mathrm{GS}^2(E_1, p) - p\big) \oplus \big(\mathrm{GS}^2(E_1, p) - p\big)^\perp,$$

where $(\mathrm{GS}^2(E_1, p) - p)^\perp$ is also a real Hilbert space as a closed subspace of the Hilbert space M_a. We assume that $\beta = \{\beta_i\}_{i \in \mathbb{N}}$ is a complete orthonormal sequence in the Hilbert space M_a such that $\beta_i = \frac{1}{a_i} e_i$ for each $i \in \Lambda$. We note that

$$\begin{aligned}
\Lambda &= \Lambda_\beta\big(\mathrm{GS}^2(E_1, p)\big) \\
&= \big\{i \in \mathbb{N} : \text{there are } z \in \mathrm{GS}^2(E_1, p) \text{ and } \alpha \in \mathbb{R} \setminus \{0\} \\
&\qquad \text{satisfying } z + \alpha\beta_i \in \mathrm{GS}^2(E_1, p)\big\}.
\end{aligned}$$

Let K be either a β-basic cylinder or a β-basic interval defined by

$$\begin{aligned}
K &= \left\{ \sum_{i=1}^{\infty} k_i \beta_i : k_i \in [0, a_i] \text{ for } i \in \Lambda;\ k_i = p_i \text{ for } i \notin \Lambda \right\} \\
&= \left\{ \sum_{i \in \Lambda} k_i \left(\frac{1}{a_i} e_i \right) + \sum_{i \notin \Lambda} p_i \beta_i : k_i \in [0, a_i] \text{ for } i \in \Lambda \right\},
\end{aligned}$$

where we set $p = \sum_{i=1}^{\infty} p_i\beta_i \in E_1$. We note that if

$$x = \sum_{i=1}^{\infty} x_i e_i = \sum_{i\in\Lambda} a_i x_i \left(\frac{1}{a_i} e_i\right) + \sum_{i\notin\Lambda} p_i\beta_i \in E_1 \subset I^\omega,$$

then $a_i x_i \in [0, a_i]$ for all $i \in \Lambda$. Hence, comparing K with Definition 4.22, we get $E_1 \subset K$. Thus, we obtain $p \in E_1 \subset K$ and $\Lambda_\beta(K) = \Lambda = \Lambda_\beta(\mathrm{GS}^2(E_1, p))$. According to Theorem 4.25 with $H = \mathrm{GS}^2(E_1, p) - p$ and $J_\beta = K$, it holds that $K \subset \mathrm{GS}(K, p) \subset \mathrm{GS}^2(E_1, p)$. Hence, we have

$$E_1 \subset K \subset \mathrm{GS}^2(E_1, p). \tag{6.5}$$

$(b.2)$ If we divide K into finitely many d_a-isometric translations $\{K_1, \ldots, K_m\}$ of a degenerate β-basic cylinder or a β-basic interval whose diameter is less than δ, then $\{K_1, K_2, \ldots, K_m\}$ is a δ-covering of E_1. Hence, considering (6.5), we obtain

$$\mu_\delta^*(E_1) \leq \sum_{i=1}^{m} \mu_\delta^*(K_i) \leq \sum_{i=1}^{m} \mathrm{vol}(K_i) = \mathrm{vol}(K) = 0$$

for any $\delta > 0$, because K is degenerate. Therefore, $\mu^*(E_1) = \lim_{\delta\to 0+} \mu_\delta^*(E_1) = 0$.

On the other hand, we see that $E_2 = f(E_1) = F_2(E_1) \subset F_2(K)$. Since F_2 is a surjective d_a-isometry and K is either a degenerate β-basic cylinder or a β-basic interval, it follows that $\mathrm{vol}(F_2(K)) = \mathrm{vol}(K) = 0$. Hence, we obtain $\mu_\delta^*(E_2) \leq \mu_\delta^*(F_2(K)) = 0$ for all $\delta > 0$. Therefore, we conclude that $\mu^*(E_1) = 0 = \mu^*(E_2)$.

(c) Now, we consider the case where $\Lambda(\mathrm{GS}^2(E_1, p)) = \mathbb{N}$, or equivalently, the case where $\mathrm{GS}^2(E_1, p) = M_a$ (see Theorem 4.26). Let $\delta > 0$ be given. By the definition of μ_δ^*, for any $\varepsilon > 0$, there exists a δ-covering $\{K_i : i \in \mathbb{N}\}$ of E_1 with cylinders from $\mathcal{C}_\delta$ such that

$$\sum_{i=1}^{\infty} \mathrm{vol}(K_i) \leq \mu_\delta^*(E_1) + \varepsilon. \tag{6.6}$$

By the definitions of $\mathcal{B}_\delta$ and $\mathcal{C}_\delta$, there exists a basic cylinder $J_i \in \mathcal{B}_\delta$ and a surjective d_a-isometry $f_i : J_i \to K_i$ for each $i \in \mathbb{N}$. Since $F_2 \circ f_i : J_i \to F_2(K_i)$ is a surjective d_a-isometry, $\mathrm{vol}(F_2(K_i)) = \mathrm{vol}(J_i)$ for all $i \in \mathbb{N}$ and $\{F_2(K_i) : i \in \mathbb{N}\} = \{(F_2 \circ f_i)(J_i) : i \in \mathbb{N}\}$ is a δ-covering of E_2 with cylinders from $\mathcal{C}_\delta$. Moreover, by applying Definition 6.6 to the surjective d_a-isometry $f_i : J_i \to K_i$, we obtain $\mathrm{vol}(J_i) = \mathrm{vol}(K_i)$. Thus, we have

$$\mathrm{vol}\big(F_2(K_i)\big) = \mathrm{vol}(K_i) \tag{6.7}$$

for all $i \in \mathbb{N}$. Therefore, it follows from (6.6) and (6.7) that

$$\mu_\delta^*(E_2) \leq \sum_{i=1}^{\infty} \text{vol}\big(F_2(K_i)\big) = \sum_{i=1}^{\infty} \text{vol}(K_i) \leq \mu_\delta^*(E_1) + \varepsilon.$$

Since we can choose a sufficiently small $\varepsilon > 0$, we conclude that $\mu_\delta^*(E_2) \leq \mu_\delta^*(E_1)$.

Conversely, if we exchange the roles of E_1 and E_2 in the previous part, then we get $\mu_\delta^*(E_1) \leq \mu_\delta^*(E_2)$. Hence, we conclude that $\mu_\delta^*(E_1) = \mu_\delta^*(E_2)$ for any $\delta > 0$. Therefore, it follows from (6.4) that $\mu^*(E_1) = \mu^*(E_2)$. □

The following lemmas are the same as the lemmas [6, Lemma 3.2, Lemma 3.3, Lemma 3.4]. It is easy to prove that $\mu^*(I^\omega) \leq 1$.

Lemma 6.9. $\mu^*(I^\omega) \leq 1$.

Proof. We apply an idea from the proof of [17, Lemma 1]. For any $\delta > 0$, there exist positive integers m and n such that $I^n = [0,1]^n$ is covered by m^n isometric n-cubes C_i which are closed in I^n with non-overlapping interiors and $d_a(J_i) < \delta$ for all $i \in \{1, 2, \ldots, m^n\}$, where J_i is the cylinder in I^ω over C_i, i.e., $J_i \in \mathcal{B}_\delta$ for each i. Then, $\{J_1, J_2, \ldots, J_{m^n}\}$ is a δ-covering of I^ω with non-degenerate basic cylinders from $\mathcal{B}_\delta$ and

$$\mu_\delta^*(I^\omega) \leq \sum_{i=1}^{m^n} \text{vol}(J_i) = 1.$$

Hence, it follows from (6.4) that $\mu^*(I^\omega) = \lim_{\delta \to 0+} \mu_\delta^*(I^\omega) \leq 1$. □

Lemma 6.10. *Given real numbers δ and b with $0 < \delta < \frac{1}{2}$ and $0 \leq b \leq 1$, define*

$$\alpha = \begin{cases} 0 & (\textit{for } 0 \leq b < \delta), \\ b - \delta & (\textit{for } \delta \leq b \leq 1) \end{cases} \quad \textit{and} \quad \beta = \begin{cases} b + \delta & (\textit{for } 0 \leq b \leq 1 - \delta), \\ 1 & (\textit{for } 1 - \delta < b \leq 1). \end{cases}$$

Then $0 < \beta - \alpha \leq 2\delta$ and $\alpha \leq b \leq \beta$.

Proof. (a) If $0 \leq b < \delta$, then $0 \leq b < \delta < 1 - \delta$ and we get $\alpha = 0$ and $\beta = b + \delta < 2\delta$. Hence, it follows that $0 < \beta - \alpha < 2\delta$ and $\alpha \leq b < \beta$.

(b) If $\delta \leq b \leq 1 - \delta$, then $\alpha = b - \delta$ and $\beta = b + \delta$. Thus, we have $0 < \beta - \alpha = 2\delta$ and $\alpha < b < \beta$.

(c) Finally, if $1 - \delta < b \leq 1$, then we see that $\alpha = b - \delta$ and $\beta = 1$. So, we have $0 < \beta - \alpha = 1 - b + \delta < 2\delta$ and $\alpha < b \leq \beta$. □

For every basic cylinder $J \in \mathcal{B}$, let ∂J denote the boundary of J. In the following lemma, we will prove that $\mu^*(\partial J) = 0$.

Lemma 6.11. *If $J \in \mathcal{B}$, then $\mu^*(\partial J) = 0$.*

Proof. Similarly as in the proof of Lemma 6.9, there exist positive integers m and n such that $I^n = [0,1]^n$ is covered by m^n isometric n-cubes C_i^n ($i \in \{1, 2, \ldots, m^n\}$) with non-overlapping interiors, where each C_i^n is a closed subset of I^n.

Let δ and b be any real numbers with $0 < \delta < \frac{1}{2}$ and $0 \le b \le 1$, α and β be defined as in Lemma 6.10, $P_i = C_i^n \times [\alpha, \beta]$ be an $(n+1)$-dimensional rectangular parallelepiped, and let J_i denote the cylinder in I^ω over P_i with $d_a(J_i) < \delta$ for all $i \in \{1, 2, \ldots, m^n\}$, i.e., $J_i \in \mathcal{B}_\delta$ for each $i \in \{1, 2, \ldots, m^n\}$.

In view of Lemma 6.10, $\{J_1, J_2, \ldots, J_{m^n}\}$ is a δ-covering of a hyper-plane H given by

$$H = \big\{(x_1, x_2, \ldots) \in I^\omega : x_{n+1} = b\big\}. \tag{6.8}$$

Hence, by the definition of μ_δ^* and using Lemma 6.10 again, it holds that

$$\mu_\delta^*(H) \le \sum_{i=1}^{m^n} \mathrm{vol}(J_i) = \sum_{i=1}^{m^n} \frac{1}{m^n}(\beta - \alpha) \le 2\delta$$

and further we get

$$\mu^*(H) = \lim_{\delta \to 0+} \mu_\delta^*(H) = 0. \tag{6.9}$$

Let J be a basic cylinder in $\mathcal{B}$. In view of Definition 4.20, without loss of generality, we will deal with the basic cylinder of the form

$$J = \big\{(x_1, x_2, \ldots) \in I^\omega : p_{1i} \le x_i \le p_{2i} \text{ for all } i \in \mathbb{N}\big\}$$

only, where there exists a positive integer n such that $0 \le p_{1i} < p_{2i} \le 1$ for each $i \in \{1, 2, \ldots, n\}$ and $p_{1i} = 0$, $p_{2i} = 1$ for all $i > n$. Then, there are at most countably many hyper-planes $H_1, H_2, \ldots$ of the forms

$$H_{2i-1} = \big\{(x_1, x_2, \ldots) \in I^\omega : x_i = p_{1i}\big\}$$

and

$$H_{2i} = \big\{(x_1, x_2, \ldots) \in I^\omega : x_i = p_{2i}\big\}$$

satisfying

$$\partial J \subset \bigcup_{k=1}^{\infty} H_k. \tag{6.10}$$

Finally, it follows from (6.9) and (6.10) that

$$\mu^*(\partial J) \le \mu^*\left(\bigcup_{k=1}^{\infty} H_k\right) \le \sum_{k=1}^{\infty} \mu^*(H_k) = 0,$$

which completes the proof. $\square$

6.5 Efficient Coverings

Let J be a basic cylinder given by Definition 4.20 that is d_a-isometric to a cylinder K via a surjective d_a-isometry $f : J \to K$. We now define

$$J^* = \prod_{i=1}^{\infty} J_i^*, \quad \text{where} \quad J_i^* = \begin{cases} [0, b^*] & (\text{for } i \in \Lambda_1), \\ [p_{1i}, p_{1i} + b^*] & (\text{for } i \in \Lambda_2 \cup \Lambda_3), \\ \{p_{1i}\} & (\text{for } i \in \Lambda_4), \\ [0, 1] & (\text{otherwise}) \end{cases} \tag{6.11}$$

and b^* is a sufficiently small positive real number in comparison with each of p_{2i}, $p_{2i} - p_{1i}$, and $1 - p_{1i}$ for all $i \in \Lambda_1$, $i \in \Lambda_2$, and $i \in \Lambda_3$, respectively. We then note that $J^* \subset J$. Taking Remark 6.5 (ii) and (iii) into account, we can cover the cylinder $K = f(J)$ with a finite number of translations of $f(J^*)$ as efficiently as we wish by choosing the b^* sufficiently small (see the illustration and Lemma 6.12 below).

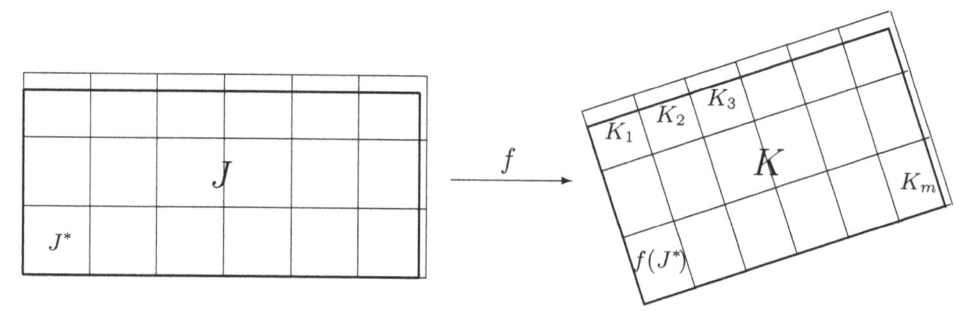

A finite number of translations of $f(J^*)$ cover $K = f(J)$

J^* is a basic cylinder and the restriction $f|_{J^*} : J^* \to f(J^*)$ is a surjective d_a-isometry. Thus, $f(J^*)$ is a cylinder, i.e., $f(J^*) \in \mathcal{C}$ (see Remark 6.2). Applying this argument, Remark 6.5 (ii), (iii) and Lemma 6.11, we obtain the following lemma that is an improved version of [6, Lemma 4.1]. This new version includes the degenerate case.

Lemma 6.12. *Let $\delta > 0$ and $\varepsilon > 0$ be given. If K is a cylinder from $\mathcal{C}_\delta$, then there exists a finite number of translations $K_1, K_2, \ldots, K_m$ of some cylinder in $\mathcal{C}_\delta$ (for example, $f(J^*)$ above and see the corresponding illustration above) such that*

(i) *$K_i \cap K_j$ $(i \neq j)$ is included in the union of at most countably many (d_a-isometric images of) hyper-planes and these hyper-planes are of the form $H_\ell^b = \{(x_1, x_2, \ldots) \in I^\omega : x_\ell = b\}$, for some $\ell \in \mathbb{N}$ and $0 \leq b \leq 1$, with $\mu^*(H_\ell^b) = 0$.*

(ii) $\{K_1, K_2, \ldots, K_m\}$ *is a covering of* K*, i.e.,* $K \subset \bigcup_{i=1}^{m} K_i$.

(iii) $\sum_{i=1}^{m} \mathrm{vol}(K_i) \leq \mathrm{vol}(K) + \varepsilon$.

Proof. We can choose a $J \in \mathcal{B}_\delta$ and a surjective d_a-isometry $f : J \to K$, where J is a basic cylinder of the form given in Definition 4.20. We now define a basic cylinder J^* by (6.11) such that $J^* \subset J$. Then J can be covered with at most m translations of the basic cylinder J^*, where we set

$$m = \prod_{j \in \Lambda_1} \left(\left[\frac{p_{2j}}{b^*} \right] + 1 \right) \times \prod_{j \in \Lambda_2} \left(\left[\frac{p_{2j} - p_{1j}}{b^*} \right] + 1 \right) \times \prod_{j \in \Lambda_3} \left(\left[\frac{1 - p_{1j}}{b^*} \right] + 1 \right)$$

and where $[x]$ denotes the largest integer not exceeding the real number x. This fact, together with Remark 6.5 (ii) and (iii), implies that the cylinder $K = f(J)$ can be covered with at most m translations of the cylinder $f(J^*)$ which are denoted by $K_1, K_2, \ldots, K_m$ (see the previous illustration).

Moreover, in view of Remark 6.5 (iii), we have

$$\begin{aligned} \mathrm{vol}(K_i) &= \mathrm{vol}(f(J^*)) = \mathrm{vol}(J^*) \\ &= \begin{cases} \prod_{j \in \Lambda_1 \cup \Lambda_2 \cup \Lambda_3} b^* & \text{(for non-degenerate } K\text{)}, \\ 0 & \text{(for degenerate } K\text{)} \end{cases} \end{aligned}$$

for each $i \in \{1, 2, \ldots, m\}$ and

$$\begin{aligned} \mathrm{vol}(K) &= \mathrm{vol}(J) \\ &= \begin{cases} \prod_{j \in \Lambda_1} p_{2j} \times \prod_{j \in \Lambda_2} (p_{2j} - p_{1j}) \times \\ \times \prod_{j \in \Lambda_3} (1 - p_{1j}) & \text{(for non-degenerate } K\text{)}, \\ 0 & \text{(for degenerate } K\text{)}. \end{cases} \end{aligned}$$

Thus, when K is a non-degenerate cylinder, we have

$$\begin{aligned}\sum_{i=1}^{m} \mathrm{vol}(K_i) &= \prod_{j\in\Lambda_1}\left(\left[\frac{p_{2j}}{b^*}\right]+1\right)\times\prod_{j\in\Lambda_2}\left(\left[\frac{p_{2j}-p_{1j}}{b^*}\right]+1\right)\times \\ &\quad\times\prod_{j\in\Lambda_3}\left(\left[\frac{1-p_{1j}}{b^*}\right]+1\right)\times \mathrm{vol}(K_1) \\ &\le \prod_{j\in\Lambda_1}\left(\frac{p_{2j}}{b^*}+1\right)\times\prod_{j\in\Lambda_2}\left(\frac{p_{2j}-p_{1j}}{b^*}+1\right)\times \\ &\quad\times\prod_{j\in\Lambda_3}\left(\frac{1-p_{1j}}{b^*}+1\right)\times\prod_{j\in\Lambda_1\cup\Lambda_2\cup\Lambda_3} b^* \\ &= \prod_{j\in\Lambda_1}(p_{2j}+b^*)\times\prod_{j\in\Lambda_2}(p_{2j}-p_{1j}+b^*)\times\prod_{j\in\Lambda_3}(1-p_{1j}+b^*) \\ &= \prod_{j\in\Lambda_1}p_{2j}\times\prod_{j\in\Lambda_2}(p_{2j}-p_{1j})\times\prod_{j\in\Lambda_3}(1-p_{1j})+O(b^*) \\ &= \mathrm{vol}(K)+O(b^*)\end{aligned}$$

and we choose a sufficiently small b^* such that the term $O(b^*)$ becomes less than ε. When K is degenerate, we have $\mathrm{vol}(K) = 0$ and $\mathrm{vol}(K_i) = 0$ for any $i \in \{1, 2, \ldots, m\}$. Hence, our assertion (iii) holds true. □

Using Lemmas 6.9 and 6.12, we will prove that $\mu^*(I^\omega) = 1$. The following theorem is equivalent to [6, Theorem 4.2], but the proof of this theorem is much more concise than that of [6, Theorem 4.2]. Hence, we will introduce the proof.

Theorem 6.13. $\mu^*(I^\omega) = 1$.

Proof. Given a $\delta > 0$ and an $\varepsilon > 0$, let $\{K_i : i \in \mathbb{N}\}$ be a δ-covering of I^ω with cylinders from $\mathcal{C}_\delta$ such that

$$\sum_{i=1}^{\infty} \mathrm{vol}(K_i) \le \mu_\delta^*(I^\omega) + \frac{\varepsilon}{2}. \tag{6.12}$$

In view of Lemma 6.12, for each K_i, there exist translations $K_{i1}, K_{i2}, \ldots, K_{im_i}$ of some cylinder in $\mathcal{C}_\delta$ such that

(i) $K_{ij} \cap K_{i\ell}$ $(j \neq \ell)$ is included in the union of at most countably many (d_a-isometric images of) hyper-planes of the form $\{(x_1, x_2, \ldots) \in I^\omega : x_\ell = b\}$, for some $\ell \in \mathbb{N}$ and $0 \le b \le 1$, whose μ^*-measures are 0.

(ii) $K_i \subset \bigcup\limits_{j=1}^{m_i} K_{ij}$.

(iii) $\sum\limits_{j=1}^{m_i} \mathrm{vol}(K_{ij}) \leq \mathrm{vol}(K_i) + \frac{\varepsilon}{2^{i+1}}$.

We notice that each K_{ij} is a cylinder from $\mathcal{C}_\delta$. If we replace the covering $\{K_i : i \in \mathbb{N}\}$ with a new δ-covering $\{K_{ij} : i \in \mathbb{N}; j \in \{1, 2, \ldots, m_i\}\}$, then it follows from (iii) that

$$\sum_{i=1}^{\infty}\sum_{j=1}^{m_i} \mathrm{vol}(K_{ij}) \leq \sum_{i=1}^{\infty} \mathrm{vol}(K_i) + \frac{\varepsilon}{2}$$

and it follows from this inequality and (6.12) that

$$1 = \mathrm{vol}(I^\omega) \leq \sum_{i=1}^{\infty}\sum_{j=1}^{m_i} \mathrm{vol}(K_{ij}) \leq \sum_{i=1}^{\infty} \mathrm{vol}(K_i) + \frac{\varepsilon}{2} \leq \mu_\delta^*(I^\omega) + \varepsilon.$$

If we take a sufficiently small value of $\varepsilon > 0$, then we have $\mu_\delta^*(I^\omega) \geq 1$ and $\mu^*(I^\omega) = \lim\limits_{\delta \to 0+} \mu_\delta^*(I^\omega) \geq 1$. On the other hand, in view of Lemma 6.9, we have $\mu^*(I^\omega) \leq 1$. Hence, we conclude that $\mu^*(I^\omega) = 1$. □

6.6 Ulam's Conjecture on Invariance of Measure

According to Theorems 3.26 and 3.30, all Borel sets in M_a are μ^*-measurable. Moreover, each Borel subset of I^ω is also a Borel subset of M_a, i.e., each Borel subset of I^ω is μ^*-measurable.

On account of Theorems 6.8 and 6.13, the outer measure μ^* is d_a-invariant with $\mu^*(I^\omega) = 1$. The proof of the following lemma is the same as that of [6, Lemma 5.1].

Lemma 6.14. *The outer measure μ^* coincides with the standard product probability measure π on the Borel subsets of I^ω.*

Proof. Similarly as in the proof of Lemma 6.9, consider a covering of $I^n = [0, 1]^n$ by m^n isometric n-cubes with non-overlapping interiors, where each n-cube is closed in I^n. Let $\mathcal{Z}_{mn}$ be the collection of cylinders in I^ω over those n-cubes, i.e., $\mathcal{Z}_{mn} \subset \mathcal{B}$. The translation-invariance of d_a implies that of μ^*. Thus, we have

$$\mu^*(J_1) = \mu^*(J_2) \tag{6.13}$$

for all $J_1, J_2 \in \mathcal{Z}_{mn}$.

Since all Borel subsets of I^ω are μ^*-measurable, by Lemma 6.11, we have

$$\mu^*(J) = \mu^*(J^\circ) \tag{6.14}$$

for any $J \in \mathcal{Z}_{mn}$ and

$$\mu^*\Bigg(\bigcup_{J \in \mathcal{Z}_{mn}} J^\circ \Bigg) \leq \mu^*(I^\omega) \leq \mu^*\Bigg(\bigcup_{J \in \mathcal{Z}_{mn}} J \Bigg).$$

We notice that $J_1^\circ \cap J_2^\circ = \emptyset$ for any distinct $J_1, J_2 \in \mathcal{Z}_{mn}$. Hence, it follows from Theorem 6.13, (6.13), and the last relations that

$$m^n \mu^*(J^\circ) \leq 1 \leq m^n \mu^*(J)$$

and it moreover follows from (6.14) that

$$\mu^*(J) = \frac{1}{m^n} = \pi(J) \tag{6.15}$$

for all $J \in \mathcal{Z}_{mn}$.

Referring to the proof of Lemma 5.21, the basic cylinders from $\bigcup_{m=1}^{\infty} \bigcup_{n=1}^{\infty} \mathcal{Z}_{mn}$, together with the empty set, generate the Borel σ-algebra over I^ω. Hence, by well known facts, (6.15) implies that μ^* coincides with π on the Borel subsets of I^ω. □

According to Theorem 6.8, assuming the axiom of choice, the outer measure μ^* is d_a-invariant. Hence, assuming the axiom of choice and using Lemma 6.14, we get the following main result:

Theorem 6.15. *For any sequence $a = \{a_i\}_{i \in \mathbb{N}}$ of positive real numbers satisfying the condition $\sum_{i=1}^{\infty} a_i^2 < \infty$, the standard product probability measure π on I^ω is d_a-invariant. More precisely, if E_1 and E_2 are arbitrary Borel subsets of I^ω that are d_a-isometric to each other, then $\pi(E_1) = \pi(E_2)$.*

Bibliography

1. E. Borel, Sur quelques points de la théorie des fonctions. Ann. École. Norm. sup (3), **12**, 9–55 (1895)

2. K.R. Davidson, A.P. Donsig, *Real Analysis and Applications* (Springer, New York, 2010)

3. L. Debnath, P. Mikusiński, *Introduction to Hilbert Spaces with Applications*, 2nd edn. (Academic Press, New York, 2005)

4. J.W. Fickett, Approximate isometries on bounded sets with an application to measure theory. Studia Math. **72**(1), 37–46 (1982)

5. S.-M. Jung, *Günstige Überdeckung von kompakten Mengen – Hausdorff-maß und Dimension*, Doctoral Thesis at University of Stuttgart, 1994

6. S.-M. Jung, The conjecture of Ulam on the invariance of measure on Hilbert cube. J. Math. Anal. Appl. **481**(2), Article 123500 (2020)

7. S.-M. Jung, Corrigendum to "The conjecture of Ulam on the invariance of measure on Hilbert cube" [J. Math. Anal. Appl. 481 (2) (2020) 123500]. J. Math. Anal. Appl. **490**(2), Article 124080 (2020)

8. S.-M. Jung, The conjecture of Ulam on the invariance of measure on Hilbert cube. http://export.arXiv.org/pdf/1807.05624

9. S.-M. Jung, Extension of isometries in real Hilbert spaces. Open Math. **20**(1), 1353–1379 (2022)

10. S.-M. Jung, E. Kim, On the conjecture of Ulam on the invariance of measure in the Hilbert cube. Colloq. Math. **152**(1), 79–95 (2018)

S.-M. Jung, *Ulam's Conjecture on Invariance of Measure in the Hilbert Cube*, Frontiers in Mathematics, https://doi.org/10.1007/978-3-031-30886-4

11. S.-M. Jung, E. Kim, Corrigendum to "On the conjecture of Ulam on the invariance of measure in the Hilbert cube" (Colloq. Math. 152 (2018), 79–95). Colloq. Math. **164**(1), 161–170 (2021)

12. S.-M. Jung, D. Nam, Some properties of interior and closure in general topology. Mathematics **7**(7), 624 (2019)

13. R.H. Kasriel, *Undergraduate Topology* (W. B. Saunders, Philadelphia, 1971)

14. P. Mankiewicz, On extension of isometries in normed linear spaces. Bull. Acad. Polon. Sci. Ser. Sci. Math. Astron. Phys. **20**, 367–371 (1972)

15. M.E. Munroe, *Introduction to Measure and Integration* (Addison-Wesley, Reading, MA, 1953)

16. J. Mycielski, Remarks on invariant measures in metric spaces. Colloq. Math. **32**, 105–112 (1974)

17. J. Mycielski, A conjecture of Ulam on the invariance of measure in Hilbert's cube. Studia Math. **60**, 1–10 (1977)

18. C.A. Rogers, *Hausdorff Measures* (Cambridge University Press, Cambridge, 1998)

19. G.F. Simmons, *Introduction to Topology and Modern Analysis* (McGraw-Hill, London, 1963)

20. S.M. Ulam, *A Collection of Mathematical Problems* (Interscience, New York, 1960)

21. S.M. Ulam, *Problems in Modern Mathematics* (Wiley, New York, 1964)

22. E.W. Weisstein, *CRC Encyclopedia of Mathematics*, 3rd edn (CRC Press, Boca Raton, 2009)

23. J.H. Wells, L.R. Williams, *Embeddings and Extensions in Analysis*. Erg. der Math., vol. 84 (Springer, Berlin/Heidelberg, 1975)

Index

Symbols

S.-M. Jung, *Ulam's Conjecture on Invariance of Measure in the Hilbert Cube*, Frontiers in Mathematics, https://doi.org/10.1007/978-3-031-30886-4

Printed in the United States
by Baker & Taylor Publisher Services

Printed in the United States
by Baker & Taylor Publisher Services